FLORIDA
for Boomers

FLORIDA
for Boomers

A GUIDE TO REAL ESTATE

Ryan Erisman

Outskirts Press, Inc.
Denver, Colorado

The author has used best efforts in preparing this book, however no warranties as to the accuracy or completeness of the information provided herein are expressed or implied. The author is not qualified to render legal, tax, or financial advice and the reader should consult with proper professionals before acting upon the information in this book. The author and publisher disclaim any liability, loss, or risk that is incurred as a consequence of the use and application of the contents of this book. The opinions expressed in this manuscript are solely the opinions of the author and do not represent the opinions or thoughts of the publisher.

FLORIDA FOR BOOMERS
A Guide to Real Estate
Copyright © 2007 Ryan Erisman
All Rights Reserved

Cover Postcard Image Courtesy of CardCow.com
All Rights Reserved. Used With Permission.
Map of Florida on page 3 is from nationalatlas.gov

Cover and book design by arismandesign.com

This book may not be reproduced, transmitted, or stored in whole or in part by any means, including graphic, electronic, or mechanical without the express written consent of the publisher except in the case of brief quotations embodied in critical articles and reviews.

Outskirts Press
http://www.outskirtspress.com

ISBN-10: 1-4327-0333-1
ISBN-13: 978-1-4327-0333-2

Outskirts Press and the "OP" logo are trademarks belonging to Outskirts Press, Inc.

Printed in the United States of America

For Kristy and Ryan Jr.

CONTENTS

ACKNOWLEDGEMENTS xiii

INTRODUCTION xv

1 AN INTRODUCTION TO FLORIDA 1

Florida Fast Facts 1
Why Do People Move to FL 2
Regions of Florida 3
MyFlorida.com 5

2 CHOOSING AN AREA 6

Getting Started 6
Get Personal 7
Government 8
Where is the Best Place to Retire in Florida 8
Home Prices 8
Home Guides 9

Magazines and Books   9
Florida Real Estate Shows   10
Contact a Real Estate Agent   10
Make a Visit   10

3 HOW TO FIND A REAL ESTATE AGENT   13

Non-Representation   14
Types of Representation   14
Special Designations   16
The Rest of Your Real Estate Team   18

4 TYPES OF HOMES IN FLORIDA   20

New Homes vs. Resale   20
Should you Rent   22
Single Family   22
Condominiums   23
Q & A with Mark Zilbert   24
Townhouses   25
Attached Villas   26
Manufactured Homes   26
Manufactured Home Communities   27

5 TYPES OF COMMUNITIES IN FLORIDA   28

Country Club   28
Active Adult/55-Plus Communities   29
Maintenance Free Lifestyle Communities   30
Resort and Club   32
Gated Communities   33

6 HOMEOWNERS AND CONDOMINIUM ASSOCIATIONS   35

Fees and Dues   35
Understanding the HOA Budget   36
Common Rules and Regulations   38

Protection of Home Values 40
Deciding If An HOA Is For You 40
Condo Association Fees 40
Special Assessments 41
Community Development Districts 41

7 PROPERTY TAXES 43

Property Appraisal Process 44
Millage Rate 44
State of Florida Homestead Exemptions 45
Other Possible Exemptions 46
Save Our Homes 47
Part Timers and Landlords Beware 48
Thinking of Downsizing? Think about this first 48
Property Tax Disclosure 48
Property Tax Disclosure Summary 49
2007 Proposed Property Tax Reform 50

8 HOMEOWNER'S INSURANCE 51

Getting Started 51
Ask Around 52
Next Option 52
The Last Resort 53
2007 Property Insurance Legislation Update 53
How to Lower your Wind Insurance Premiums 54
Types of Coverage 55
Inventory 55
Flood Insurance 56
Flood Zones Explained 56
Better Safe than Sorry 57

9 CONTRACTS AND DISCLOSURES 58

Demystifying the Florida "FAR/BAR" Contract 58
"FAR/BAR" Standards for Real Estate Transactions 61
Common Florida Real Estate Disclosures 65

10 HOME INSPECTIONS AND WARRANTIES 71

Home Inspections 71
Home Warranties 72

11 NEGOTIATING 74

It's Not Just About Price 75
Top Four Tips For Negotiating 75

12 FINANCING 77

Types of Lenders 77
Secondary Mortgage Market 79
Types of Mortgages 79
Special Financing 81
Interest Rates 83
Applying for a Mortgage 84
Good Faith Estimate 85
1031 Exchange 85
Your Credit Score 86
Your Credit Report 87
Mortgage Calculators 87

13 NEW HOME CONSTRUCTION 88

Choosing a Builder 88
Working with the Builder and His Staff 89
How to Choose a Floor Plan 90
Choosing a Lot 91
What is a Zero Lot Line 92
Be Aware of these Builder Contract Clauses 92
Making Your Selections 95
Critical Steps in the Florida
   New Home Construction Process 96

Contents   xi

14 YOUR NEW HOME WALKTHROUGH   125

    Allow Enough Time   125
    What to Bring   125
    Breaker Box and Electrical System   126
    Hot Water Heater   126
    Water Shutoff   127
    Air Handler and Air Filter   127
    Garage Door   127
    Kitchen   127
    Appliances   128
    Drywall and Flooring   129
    Systems and Components   129
    Bathrooms   129
    Exterior   129
    Warranty   130
    Emergency Information   130
    Sign Here Please   131

15 SWIMMING POOL CONSTRUCTION   132

16 REAL ESTATE CLOSING   136

    Title Insurance Companies   136
    Attorney's Offices   137
    Closing Costs   137
    How Should You Take Title   138

17 MAINTAINING YOUR NEW HOME   140

    Periodic Maintenance   140
    Pest Control   141

18 FLORIDA RESOURCES   142

    Population and Growth   142
    Florida's Top Ten Metropolitan Statistical Areas   143

Hurricanes   143
Getting Around Florida   146
Employment   150
Community Service   150
Continuing Education   151
Health Care   152
Arts and Cultural   154
Museums   154
Theme Parks   156
Festivals   160
Parks   161
Beaches   162
Lighthouses   162
Fishing and Boating   163
Professional Sports   163

GLOSSARY   165

WEBSITE INDEX   175

RECOMMENDED READING   183

BIBLIOGRAPHY   185

ACKNOWLEDGEMENTS

Special thanks to my wife Kristy for her support over the last year and a half that it took me to get all this down on paper. Thank you also to Ryan Jr. for letting dad get a little work done here and there.

I owe a big debt of gratitude to all of my baby boomer customers whose Florida home buying and retirement experiences were my inspiration for writing this book.

This book wouldn't be the complete resource it is without the input and editing of my sister, Brandie Erisman.

Thanks to Judith Arisman, of Arisman Design Studio for helping guide and refine my vision for the cover design. You've been a bigger help than you know.

Last but certainly not least, special thanks to my mother Marty Erisman, who since I was in the third grade has encouraged me to be a writer.

INTRODUCTION

If you are one of the 6.5 million baby boomers thinking of retiring to Florida, or at least considering buying a second home here in the next 10-15 years, this book is for you. The idea for this book was born out of frustration. For the past several years, as a real estate broker and more specifically a new home salesperson in Florida, I have worked primarily with baby boomers who were either purchasing their future retirement homes, or buying second homes to visit and enjoy with friends and family.

No matter how successful, well-educated, or bright these baby boomers were, each had their own set of questions that they needed answered about Florida. Some had basic questions such as where to retire in Florida, what each region of Florida offers, and what the weather would be like at different times of the year. Others pondered what they might do to fill their days, wondering whether they would continue to work in retirement, start new businesses, or split their days between the golf course and the beach.

Invariably, their questions would eventually turn to real estate. They needed to know things like where could they find the best values in home prices, how much would their taxes be, would they be able to get insurance and how much would the insurance cost. Others wondered what type of community would be best for them, for example, would they want to live somewhere that has a

Homeowners' Association? Many other boomers were interested in learning more about the process of having a new home built in Florida.

What was frustrating for me was seeing so many of them receive and act on misinformation that they found on an unreliable website, or that they received from well-intentioned but misinformed friends or family, and in a few rare cases, from agents or builders with only their own interests in mind. My goal for this book is to answer the questions that boomers have about buying or building a Florida home and eventually moving to Florida to enjoy their retirement.

Another problem I hope to address is that of outdated information. The real estate landscape in Florida changes frequently, and left unchanged this book would be obsolete in two to three years. My goal is to update *Florida for Boomers* every two or three years so that boomers like you will always have the most current information about real estate in Florida.

But don't worry, you won't have to spend $20 to purchase a new copy each time the book is updated. Just go now to the book's website at www.floridaforboomers.com and click on "Register Your Copy." You will be asked some basic questions, among them your e-mail address, your name, where you bought the book, and your order number or other identifying information. Once you successfully register your copy, every time the book is updated you will receive an electronic copy by email at no charge.

In addition to the information found in the hard copy of this book, the book's website at www.floridaforboomers.com will also prove to be a valuable resource for you. I have put together a password-protected "Florida Resources" page exclusively for you as a reader of this book.

Your special password to access this page is: **boomer1946**

I'm sure that you are itching to start learning all you need to know about Florida real estate, but please know one thing. The success of this book and future editions of it depends on you, the reader. Assuming you find this book to be a great resource, let other boomers know about it. The best way to go about doing this is to write a five-star review about it on the book's page at Amazon.com.

Best wishes for your Florida retirement!

1.
AN INTRODUCTION TO FLORIDA

So, you're thinking of retiring to Florida or maybe buying a second home or vacation home here. You won't exactly be blazing a new trail. Though likely inhabited by Native Americans for thousands of years before Europeans set eyes on it, Florida was "discovered" in 1513 by Juan Ponce de Leon who claimed it for Spain. Ponce de Leon named it Florida, because he landed here during "Pascua Florida," the festival of flowers during the Easter season.

The next several hundred years were a tumultuous time in Florida with various parts of the state changing hands several times between the Spanish, French, and British. After Britain's defeat in the Revolutionary War, Spain regained control of Florida, but later ceded it to the United States. Florida became the 27th state on March 3, 1845.

Florida Fast Facts

Capital: Tallahassee
Nickname: The Sunshine State
State Flower: Orange Blossom
State Beverage: Orange Juice
State Bird: Mockingbird
State Tree: Sabal Palmetto Palm
Funky Fact: State Pie: Key Lime

Since it became a state, Florida has seen steady and sometimes explosive growth. Today more than a thousand people a day set their sights on moving to Florida. Most of them are chasing fantastic year-round weather and a slower pace of life.

WHY DO PEOPLE MOVE TO FLORIDA?

The two main reasons people consistently name for moving to Florida are the great weather, and the slower pace of life. But let me tell you about another attractive feature of living in Florida.

Taxes

Florida is consistently ranked among the states with the lowest tax rate. It has less than half the tax burden per capita as New Jersey, the nation's highest tax state per capita, making it a haven for retirees as well as businesses. The state sales tax is 6% and the corporate tax is 5.5%. One thing that attracts many people to Florida is that there is no state income tax. Any income that you plan to pull in from investments, pensions, or a second career during your retirement will only be taxed at the federal rate, which means more money for you in the end.

Another advantage is that since January 1, 2007, Florida residents no longer have to file an intangibles tax return on their stocks, bonds, mutual funds, money market accounts and other investments.

Florida Weather

Florida is well known for it's beautiful weather, with plenty of sunshine most days of the year. Thousands of Northerners affectionately referred to as "Snow Birds" flock to the state each winter to escape the frigid winter temperatures up north.

During the winter, temperatures average around the mid-fifties in the north part of the state, and the mid-sixties down south. You won't find coat closets in most of the homes you see in Florida, but don't let that fool you. In some parts of the state it can get below freezing, and it has been known to snow.

Florida's weather is primarily subtropical, largely because it is nearly surrounded by water. In the summertime temperatures can get uncomfortably warm. Lots of folks who move to Florida from up north use the summer months to do some traveling, going back

AN INTRODUCTION TO FLORIDA

up north to see friends or traveling abroad. This is something that you may want to consider as well. There are days, especially in South Florida, where the mercury can top 100 degrees.

The average temperature in North Florida during the summertime is around 80 degrees while the average in South Florida is in the low to mid-eighties. You can also usually count on an afternoon rain shower or thunderstorm to cool you off a little bit on most summer days. Be careful during those afternoon thunderstorms, as Central Florida is known as the lightning capital of the world. Florida has a rainy season that runs from June through October. This rainy season accounts for around 70% of Florida's annual rainfall, which is between 50 and 60 inches for most parts of the state.

REGIONS OF FLORIDA

This overview of Florida admittedly does not do any of the regions justice. This section has been kept brief to make room in the book

for the important information you will need to know about Florida real estate. There are other books available for those wanting to study Florida's regions in depth. See the "Recommended Reading" section in the back of the book for some ideas.

Panhandle

This region is home to the state's capital, Tallahassee, as well as Pensacola, Fort Walton Beach, and Panama City. People love the tranquility of the Panhandle, as it is one of the least heavily developed parts of Florida. However, as more people discover this region, this tranquility may not last.

North East

The areas of Jacksonville, St. Augustine, and Palm Coast are where the bulk of the population in this fast-growing region live. Palm Coast is located in the nation's fastest growing county, Flagler County.

East Coast

Known as the "Space Coast" because shuttle launches take place here from Cape Canaveral, this region includes Ormond Beach, Daytona Beach, Port Orange, New Smyrna, Titusville, Melbourne, and Cocoa Beach. This region is growing rapidly not only because of Northerners retiring there, but also because of people moving up from South Florida, to escape the high home prices and congestion.

West Coast

Tampa, St. Petersburg, Clearwater, Bradenton, and Sarasota are the biggest population centers in this region of the state. Also known as the Gold Coast, the west coast of Florida offers direct access to the Gulf of Mexico, which is perhaps a more serene option than the Atlantic Ocean.

Central Florida

This part of Florida is best known as a tourist destination because most of the theme parks in the state are located in central Florida. Some prominent cities include Orlando, Lake Mary, Altamonte Springs, Winter Park, Kissimmee, Leesburg, Ocala, Gainesville, and Lakeland.

South West

This region is popular with people from the Midwest U.S. who can hop on I-75 and shoot right down. Popular destinations in Southwest Florida include Naples, Marco Island, Cape Coral, Port Charlotte, and Fort Myers. Home prices in some of these cities are known to be among the highest in the state, but the beauty of the area may be unrivaled.

South East

Perhaps best known for Miami and South Beach, South East Florida offers much more. It has long been a retirement haven for people from the Northeast U.S. Home prices in this region can be high, depending on the area. Major cities include Miami, Ft. Lauderdale, Hollywood, West Palm Beach, Boca Raton, Delray Beach, and Stuart. Most people consider the Florida Keys to be a part of South East Florida as well.

MYFLORIDA.COM WEBSITE

The state of Florida has compiled a great resource for residents and visitors at its official website, myflorida.com. It's a portal for almost any kind of Florida information you would want or need. Seeking information about how to get a Florida driver's license? Myflorida.com. Want to research the history of Florida? Myflorida.com. Looking for data about taxes, insurance, roadways, employment, recreation, or health care in Florida? Myflorida.com. You can even write a letter to the governor (Charlie Christ, who was inaugurated on January 2, 2007) from the website. Anytime you are not sure where to look for information about Florida, start at myflorida.com.

Visit http://www.myflorida.com

2.
CHOOSING AN AREA

Choosing an area in which to live in Florida can be difficult, especially if you have not spent a lot of time in Florida yet. There are many great choices available to you. But you can use these tips to narrow your choices down and eventually make a final decision on a place that fits your lifestyle and budget, suits your housing needs and desires, and that you'll love for years to come.

GETTING STARTED

Since you likely live hundreds—if not a thousand—miles away from Florida, the best place to start your search is online. Visit the websites of the local newspapers for the cities, towns, or areas you are investigating and do some general exploring and reading. You can often find information on the history of an area, photo tours and sometimes virtual tours. Don't forget to scan the real estate classifieds to get a feel for home prices in the area. It will also be helpful to read some local articles and editorials. These can give you a sense of the feel of an area, and reveal items of interest or concern for the local residents.

> Links to most major newspapers in Florida can be found at http://www.floridaforboomers.com then click on "Florida

Resources". Again, your special password to access this page is: boomer1946

Radio and TV news stations also have some excellent resources on their websites. Radio stations can be especially helpful for finding information in line with your interests. For example, a community calendar for a country music station might list events that would appeal to their typical listener. These events would be different than events that might interest an oldies or easy listening crowd.

Links to several Florida radio stations and TV news stations can be found at http://www.floridaforboomers.com, then click on "Florida Resources".

Next, visit the websites of the area chambers of commerce and request an information package. The information packages will usually include a brochure on the area filled with advertisements for local businesses, information on annual events, a guide to local history, and usually a map of the area that will come in handy when you make an in person visit. In addition, request information from some new home builders and communities in the area, to see if the types of houses they offer appeal to your wants and needs. If the area seems like it might meet your requirements, go to the next step.

Links to most Chambers of Commerce in Florida can be found at http://www.floridaforboomers.com then click on "Florida Resources"

GETTING PERSONAL

It goes without saying that you should talk to friends or relatives living in Florida, but don't overlook people in your extended network or sphere of influence. For example, did your golf buddy's brother-in-law recently move to Florida? Give him a call and get the scoop.

Visit online groups and search for topics involving "Florida," "retirement," and "real estate." Google Groups and Yahoo Groups are great starting places. You can usually count on people being quite frank with their feelings about each topic in these online groups.

Check out: http://groups.google.com and http://groups.yahoo.com

GOVERNMENT

Visit the website of the local government. By the looks of things, are they up-to-date technologically? Read through official statements and press releases you may find. Does the local government seem prepared to manage the growth their area may be experiencing?

Also, what are the taxes like in the area? Is the government being wise and prudent in their spending, or does it seem like they are plundering windfalls from property tax increases? These are all not easily answered questions but with a little research you can get a feel for what some of the answers might be.

WHERE IS THE BEST PLACE TO RETIRE IN FLORIDA?

It's impossible to say exactly where "the" best place in Florida is to retire. It should be different for everyone. But Money Magazine has compiled data on more than 50 areas in Florida, allowing you to search for your ideal retirement location based on different criteria such as availability of health care, affordable housing, recreation, amount of population over 50, and other criteria you select.

Visit: http://money.cnn.com/magazines/moneymag/bpretire/2006/states/FL.html

HOME PRICES

To get a general feel for what the prices of homes are in the area in which you are looking, visit sites like Realtor.com. They will show you all the homes listed in the Multiple Listing Service for the city or zip code that you specify. Enter your required number of bedrooms and bathrooms as well as your price range and see which homes come up as a result.

Two real estate websites, Zillow.com and Trulia.com, have neat features where you can enter a zip code or city and see a color-coded map which reveals where the prices are more or less expensive.

http://www.realtor.com
http://www.zillow.com
http://www.trulia.com

CHOOSING AN AREA

HOME GUIDES

There are several real estate and home guides available that can be a good resource for finding real estate in Florida. Guides such as *Homes and Land*, *The Real Estate Book*, and *Digest of Homes* can be ordered by phone or on-line for the specific areas you are interested in. These magazines contain ads for available homes and for real estate agents and lenders promoting their services.

One caution: If you sign up by phone or on-line for a magazine to be mailed to you, they will of course ask for your address so they can send the magazine. They will also ask you for your email and phone number, which they do not need to send you the magazine. Your contact information is then distributed in many cases to the agents and lenders advertising in these magazines. If you do not want anyone to bother you with e-mails or phone calls, keep your phone number and e-mail address confidential.

For more information or to order magazines on-line:
http://www.homesandland.com
http://www.therealestatebook.com
http://www.digestofhomes.com

MAGAZINES AND BOOKS

There are three magazines that you'll want to check out for their informative content as well as community listings and advertisements. These magazines include *Where to Retire Magazine* (wheretoretire.com), *Living Southern Style* (livesouth.com), and *2nd Home Journal* (2ndhome.net). Each has its own style and while none focus exclusively on Florida, there are plenty of advertisers from Florida communities on the pages of each.

Another helpful resource is the book *Where to Retire in Florida*, by Richard and Betty Fox (Vacation Publications, 1999). The book rates 99 cities and towns in Florida based on specific criteria such as crime, taxes, jobs, recreation, and more. Because it was published back in 1999, you should use the information in the book as a guideline only. Places change so quickly, and you'll be best served by doing your own research. Nonetheless, *Where to Retire in Florida* is a great starting point for that research.

Last but not least, to round out your retirement bookshelf, pick up a copy of *The New Retirement* by Jan Cullinane and Cathy Fitzgerald (Rodale, 2004). This book covers a wide spectrum of retirement topics. While the book's content is not just limited to real estate, it does include a section on recommended locations throughout Florida to move to for retirement. The retirement planning forms and worksheets provided in the back of the book alone are worth the price of the book. The second edition of *The New Retirement* is due out around July of 2007.

FLORIDA REAL ESTATE SHOWS

Florida real estate shows are a great opportunity to check out some of the new communities in Florida without even having to set foot in the state. Home builders, developers, and real estate agents come to these shows in hopes of making a good first impression on potential buyers like you. The shows are set up so that you can wander around and stop for information at booths that look like they may be of interest to you.

Two companies that organize these shows are Live South Shows (run by the same company as the previously mentioned Living Southern Style Magazine), and Florida Lifestyle Expo. Show locations are primarily in the Northeast and Midwest U.S.

> For more information on show dates and locations visit http://www.livesouthshows.com and http://www.floridaliving.org

CONTACT A REAL ESTATE AGENT

Working with a real estate agent is discussed in depth in the next chapter, but for starters, you can investigate some agent websites in the area that you are considering. Some agents offer relocation packages that provide a lot of good information on the areas they service.

> Start your real estate agent search and screening process at http://www.floridaforboomers.com and click the "Find an Agent" button.

MAKE A VISIT

Now that you've gathered a bunch of information on places you think might interest you, its time to make a visit. When you visit an area, especially one you've never been to before, there are certain things you'll want to look for to help you decide if this is an area you might like to live in. Using the map that came in your package from the local chamber of commerce (otherwise you should buy a local map), drive through some different parts of the community.

As you are driving, are you seeing restaurants, businesses, shopping centers and so forth that look appealing to you and that match your ideal lifestyle? Stop in at some different establishments like restaurants or shopping malls and take a look around. Ask people you come in contact with what they like or don't like about the area. See if they have any recommendations of places to look for homes or any other relevant information they are willing to share.

Keep in mind that every city or town has its skid row. The areas around airports tend to be especially bad, so don't turn around and get back on a plane if you haven't ventured more than a few miles from the airport.

Drive Through Some Neighborhoods

Take a detour off the main roads and into some residential neighborhoods. Are the homes what you expected? Are people's yards well maintained? Visit some of the new communities and model homes in the area or take a tour of some resale homes that you've arranged to see ahead of time with your real estate agent.

Facilities and Services

Look for the amenities that are important to you. Are there libraries nearby and are they up to date? What about medical facilities? Check to make sure an area's cultural activities, recreation facilities, beaches and parks, golf courses, and whatever else is important to you are available at a level that will fit your needs and desires.

Make a list of your current weekly activities (garden club, rotary, church, etc.) and be sure that the area you choose provides you the opportunities to continue to do what you enjoy.

You've Got Visitors

Keep in mind the likes and dislikes of friends and family members who will be coming to visit. Do your kids and grandkids love the beach, or is a ten-minute drive to Disney World more important? Presumably you will want them to visit as often as possible, so get their input before deciding on a place.

One of the worst things that could happen to you during this transition to life in Florida is you find your dream home, but hate to go out into the community surrounding it for lack of things to do and be a part of. Or you buy your dream home, but find that when the kids and grandkids come to visit there's nothing fun for them to do. But this won't happen to you as long as you do some prudent research and investigating before you make a purchase and settle in.

3.
HOW TO FIND A REAL ESTATE AGENT

If you took the initiative to purchase and read this book, you know better than to simply trust that the first real estate agent who crosses your path will be able to adequately handle the complex details of your real estate purchase in Florida. Unfortunately, this is not the case for many people. According to the National Association of REALTORS®, 70 percent of people complete a real estate transaction with the first agent they made contact with.

In order to find the very best real estate agent for you, first ask friends or family who have moved to the area you are considering for referrals. If you don't have anyone to contact in the area you are considering, try doing an internet search, such as "Orlando real estate agent," and investigate the websites of several agents that come up. Don't forget to check out the "Find an Agent" page at floridaforboomers.com.

Also, try contacting the chamber of commerce for the city you plan to move to and see if they have any recommendations. It is a good idea to contact at least a few agents in the early stages to get a feel for what to expect down the road. Be open and forthright with them if you have used the services of another agent before contacting them, so that there are no surprises for anyone later.

Ask the agents questions about their qualifications, years of experience in their market, and whether they have helped other people, particularly boomers like you, relocate to their area. They should be able to offer you written testimonials from satisfied clients. Also, since you will be new to the area, make sure the real estate agent has a strong network of local service providers such as attorneys, home inspectors, and lenders to recommend to you.

You may want to go one step further in the screening process and see if there are any complaints against the agent. Simply go to http://myfloridalicense.com and under "Public Services" click on "Search for a License, Permit, or Registration." Then enter in the agent's name and click search. Scroll down for the results. You can also do this for any builder or contractor you are thinking of hiring.

NON-REPRESENTATION

At your first substantial meeting with a real estate agent, whether in their office, in your home, or looking at property, the agent is required by law to provide you with a No Brokerage Relationship Disclosure. This disclosure, which the agent will ask you to sign (although you don't have to if you don't want to), tells you in no uncertain terms that the agent you are speaking with or meeting with has no brokerage relationship with you. Because of this, you should not disclose to them any information you want held in confidence. With this type of relationship, the agent owes you no fiduciary responsibility.

Even though the agent does not owe you confidentiality, they do by law still owe you three duties. 1. To deal with you honestly and fairly. 2. To disclose all known facts that materially affects the value of property but may not be readily observable to you. (i.e. there used to be a chemical plant on the site where you are purchasing a home) and 3. To account for all funds that you may give them.

TYPES OF REPRESENTATION IN FLORIDA

If you should decide to allow the agent you are meeting with to represent you, you can have them represent you in a couple of different ways. While there is no one right way to have an agent represent you, it's still very important to know who is and who is not on your side in any transaction.

Single Agent

The first form of representation for you to consider is a single agent relationship. In this relationship the agent is solely on your side, that is, they owe a fiduciary duty to you. A single agent owes you the following nine duties: 1. To deal with you honestly and fairly. 2. Loyalty. 3. Confidentiality. 4. Obedience (within the scope of the law). 5. Full disclosure. 6. Accounting for all funds. 7. To use skill, care and diligence. 8. To present all offers and counteroffers in a timely manner unless otherwise instructed in writing. 9. To disclose all known facts about the property materially affecting the value of the property.

Transaction Broker

The second way for an agent to represent you is as a transaction broker. This is common if you are working with an agent who also represents the sellers of homes you are looking at. In this case, the agent is supposed to be a neutral party in the transaction. A transaction broker owes you six duties, as opposed to nine of the single agent. The main changes are the elimination of loyalty, obedience, and full disclosure. The duty of confidentiality becomes "limited confidentiality" (i.e. the agent cannot disclose to you that the seller will take a lower price, and likewise they cannot tell the seller that you are willing to pay more).

It should be noted that beginning in July 2007, it will be automatically assumed that you are dealing with a transaction broker, unless you specifically and explicitly choose to enter into a single agent agreement. But until that time, you must choose.

Buyer Brokerage

Buyer Brokerage is another type of relationship that you could engage in. The Exclusive Buyer Brokerage Agreement, if presented to you, is a document that essentially binds you to a certain agent, for a certain amount of time, to look for a certain type of property.

The thing to be careful of is that the agreement is usually written with you, the buyer, being held responsible for paying the agent's fee if the seller does not agree to pay it, or paying for any shortfall in the amount required by the agent. It should be noted here that not all buyer's agents operate in this fashion, but several do, as the Florida Association of Realtors (FAR) Exclusive Buyer Brokerage Agreement includes this compensation clause.

I'm all for working with a buyer's agent, someone who has only your interests, not the sellers, at heart. However, because compensation is usually offered through the seller, you should request that they delete the clause requiring you to pay them their fee if the seller will not. The only exception should be in the case of the agent finding you a for-sale-by-owner property that does not offer the agent compensation.

If you wish to be shown for-sale-by-owner properties, be clear up front with the agent about how much you are willing to pay. Most agents will ask for a percentage of the sale price, usually two to three percent, but you're better off paying them a flat fee to help you. For example, if you agree to pay a percentage, the agent makes out better when you pay $300k for a house than they would have if they had helped you negotiate a sales price of $270k.

SPECIAL DESIGNATIONS

As you search for an agent, you may find that the initials after their names are beginning to look a little like alphabet soup. John Smith, GRI, ABR, MBA. Jane Smith, CRB, CRS, e-Pro. It can be helpful to know what some of these letters (which abbreviate different designations that the agent has earned through experience or education) stand for. If you see an agent who has letters after their name that I haven't mentioned, just ask what the initials stand for. I'm sure the agent would be more than happy to tell you about it.

Graduate Realtor Institute – GRI

GRIs have obtained their designation by attending a minimum of 90 hours of classroom instruction on topics including contract law, professional standards, sales and marketing, finance, and risk reduction. When working with a REALTOR® who has earned the GRI designation, you are working with someone who has shown that they are dedicated to their profession, knowledgeable, and adequately trained to help your transaction go as smoothly as possible.

Council of Residential Specialists – CRS

Before obtaining the CRS designation, REALTORS® must have a considerable amount of experience, have conducted a certain volume of real estate deals, and must have completed rigorous educational requirements. Less than four percent of all REALTORS® are members of the Council of Residential Specialists, making it a very elite group of agents.

Certified New Home Salesperson – CSP

This is a designation that you might see when shopping for new construction homes. The CSP designation is offered to new home salespeople through the National Association of Home Builders (NAHB). Salespeople with the CSP designation are trained to relate to their new-home seeking customer's wants, needs, and desires, and to guide them smoothly through the sometimes rocky road that is new construction.

Accredited Buyer Representative – ABR

ABR stands for Accredited Buyer Representative. An agent with this designation has been specifically trained to work with different types of buyers, such as buyers of new homes, e-buyers, and relocation buyers. As with most of the other designations I've mentioned, the ABR designation is earned by completing extensive education requirements and possessing a proven track record of working with buyers.

Resort and Second Home Specialist

A REALTOR® with a Resort and Second Home Specialist designation can successfully guide you in your purchase of a second home or investment property in resort towns and communities. Training for this designation includes courses on creating wealth through investment real estate, managing second home and investment properties, and the essentials of international real estate. They are also trained in 1031 tax-deferred exchanges, also called "like kind" exchanges, where you can avoid paying taxes on the sale of an investment property by using the profit to purchase a property of equal or greater value. This is becoming a popular designation due to the rapid rise in second and vacation home ownership.

E-pro

If you prefer to do a large part of your home shopping online, or at least your initial research, and you would rather converse with and vet an agent through e-mail versus over the phone, then working with an e-pro may be for you. E-pros are trained to answer e-mails promptly and professionally, offer extensive property listings and other information to you electronically, all the while respecting your time and your privacy. While any REALTOR® should be able to send you pictures by e-mail of properties you may be interested in, e-pros can do that and much more. E-pros have taken the extra steps to acquire the training and expertise that will benefit the internet consumers of today.

THE REST OF YOUR REAL ESTATE TEAM

In addition to your REALTOR®, you may want to have one or more of the following people on your side during your search for, and purchase of, a new home.

Attorneys

In several areas, perhaps the place you currently live, attorneys are required in all real estate transactions. Not so here in Florida. However, this does not mean you should not use one, especially if it is a complex transaction, or there are clauses in the contract that you do not understand. It can certainly make you feel more comfortable having a set of trained eyes on any contract you sign. One point I would like to make, however, is that if you are buying real estate in Florida and would like the counsel of an attorney, use a Florida-based attorney, preferably one who specializes in real estate.

I'm sure the attorneys where you are currently living are very competent and would never purposely misguide you, but for your protection, it's better to use an attorney who deals with real estate transactions in Florida on a daily basis. You wouldn't use a medical malpractice attorney to defend you in a DUI case, would you? Then don't do the same with your real estate transaction. The best way to find a real estate attorney is through a referral from your real estate agent. They should give you a list of more than one attorney to check out on your own.

Tax Advisor and Accountant

When deciding to purchase a home in Florida, there are a number of tax considerations you should discuss with your tax advisor or accountant. Among the many concerns is whether you should pay cash for your new home or get a mortgage so that you can get the tax deduction on the interest. This is not an easily answered question, and the answer will not be the same for everyone. But your tax advisor or accountant should be able to help you decide what's best for you and your situation.

Financial Advisors

Buying real estate anywhere is both a financial and an emotional decision. Naturally, you will seek the guidance of as many people around you as you can. For some people, this includes seeking the guidance of their financial advisor.

Most financial advisors put their clients' interests before their own and should be highly regarded for the contribution they can make in developing an individual's financial strength. However, just as in any other industry, the financial advisor industry does contain some unscrupulous people who have a tendency to look out for their own interests, rather than those of the client's. If you have done all the research you can do regarding your decision to purchase a home and feel comfortable with it financially, you should not let the advice of a financial advisor or stockbroker stand in your way.

When consulting with your advisor or broker about a real estate purchase, remember that they are usually compensated based on how much money they are directing for you. If they could stand to lose hundreds of thousands of dollars from your portfolio because you want to sell some investments to pay cash for a new home, they may not be happy about parting with your money. If you sense any personal motive in their advice to you on this matter, take your money and run, don't walk, to find another advisor.

4.
TYPES OF HOMES IN FLORIDA

NEW HOMES VS. RESALE

When considering a home purchase in Florida, or anywhere for that matter, one of the first decisions you should make is whether to have a home built, or buy a resale (previously occupied) home. Your decision will depend on several factors including how quickly you need a home, your personal taste, and other factors. Here are some pros and cons of both.

Pros of Building a New Home

One of the best things about a brand new home is that it is under warranty from the builder. If (almost) anything goes wrong while the home is under warranty, you won't be charged to have it fixed. (Read more about new home warranties in Chapter 14).

Assuming you are building from scratch (not buying a builder spec home), you will get to choose your own décor like carpet, tile, cabinets and counters. This helps to personalize the home to your tastes and to give it some of your own soul. You also have the ability to customize to an extent, depending on what types of changes the builder allows.

Also, you will qualify for better insurance rates because the home will be built to current building codes.

Cons of Building a New Home

One of the cons of building a new home is that you typically have to wait for the home to be built, unless the builder has the style of home you want in his inventory (commonly referred to as "spec" homes or "quick move-in" homes). If you are on a tight schedule, or you do not want to find a temporary place to live while your home is under construction, you might want to pass on building a new home.

Also, builders are typically not negotiable on their prices, mainly because when they price their homes they use a set profit margin and hate to stray from that figure. Your best shot at getting any wiggle room on the price is with negotiating option prices. Builders usually have huge markups of 50-100% or more on options and extras not normally included in their "standard" houses.

Keep in mind that some builders are willing to negotiate on the overall price of the house, and that it also depends on the market. If they have a lot of homes in inventory (buyer's market), they may be more likely to negotiate with you than if they don't have very many homes available (seller's market).

Another important factor to consider is that building a new home can be an overwhelming and nerve-wracking process. Seeing little day-to-day progress can be exasperating and many people feel the urge to micromanage the builder when there is usually no need to do that. If you are predisposed to being a micromanager, skip the headaches and buy a home that's already built.

Pros of Resale Homes

One pro of buying a resale home is that unless you have plans to do some remodeling before you move in, the home is ready to be occupied, and you know exactly what you're getting. You get to avoid the roller coaster of emotions involved in building a new home.

Most home sellers are open to negotiation on price. Again, this depends on the market, and in some cases why the seller is selling. Helping you negotiate is where your real estate agent comes in handy.

Cons of Resale Homes

With a resale home you are not able to choose your décor such as tile and carpet, cabinets and countertops, or make any customization or personalization until after the purchase and, even then, not without a remodeling budget. It is what it is. Someone else has chosen the colors and materials, and their tastes may differ from your own. Something else to consider is that, depending on the age and construction of the home, your insurance may cost more.

Additionally, if you want the protection of a home warranty, it must be purchased separately at your expense, unless the seller provides one. Also, don't forget you'll need a home inspection, which you can read more about in Chapter 10.

SHOULD YOU RENT?

Another possibility for you to consider is renting a home for a year or two while you acclimate yourself to Florida. This can be helpful for a boomer who isn't sure they want to retire in a certain place, or live in a certain neighborhood, or type of home. You can try it out, and when your lease is up, you can decide what to do from there.

Most experts agree that in most cases buying is better than renting. Not so much because of future appreciation that can take several years to realize, but for the tax benefits of owning, like deducting mortgage interest and real estate taxes. But in a situation where you are the least bit unsure of your decision, renting may be the answer. It would do you very little good to plunk down tons of money on a new home, only to decide you hate the area, and in six months or a year later pick up and find a new home in a new area.

There are other problems with renting as opposed to buying, including the difficulty of finding something that suits your tastes and housing needs. There's no such thing as a custom built rental. Plus, if you do find a place, since you don't own the home you'll be restricted in making any changes to its appearance.

SINGLE FAMILY HOMES

The most basic and most popular type of home is the single family home. It's what most people think of when someone says "house". A standalone structure, a single family home sits on its own piece of land, be it the size of a credit card or several acres. Single family homes offer their owners the most sense of space. Even if your neighbor's home is only five feet away, as will be the case in some communities, you still have a feeling of separation and distance from them. When standing in your living room, you really can't tell if the neighbor's house is five feet or 50 feet away.

Single family homes typically offer the most flexibility when you wish to make changes, such as adding an addition, changing the exterior color, or putting in a pool.

If you buy a single family home in a subdivision governed by a Homeowners' Association (HOA)(see chapter 6), you will not have as much flexibility with what you can do to your home. The Architecture Review Board or ARB must typically approve most changes, especially those affecting the exterior appearance of the home. However the upside is that your neighbors will have to conform to the same standards when they wish to make changes. Be sure to read the HOA restrictions before purchasing to make sure they're rules you are willing to follow.

As an owner of a single family home, you will be responsible for the home's maintenance. You will be responsible for cutting the grass, trimming the shrubs and bushes, painting, pressure cleaning, and any other exterior maintenance as needed.

For the boomer who has better things to do than to spend Saturday on yard work, however, the new trend in some communities in Florida is for single-family homes to be maintained on the outside, just like a townhouse or condo. These are called "maintenance-free communities" or "maintenance-free lifestyle communities." Just as in a townhouse or condo, the owner is assessed a fee to pay for certain services such as lawn care, periodic painting, and pressure washing. You'll find more information on maintenance-free communities in Chapter 5.

CONDOMINIUMS

Condominiums, or condos, are popular all over the state, but even more so in coastal areas. Condominiums are buildings comprised of several separate units. Theoretically, the price of the land that the condo is built on is spread across the units, with units on higher floors typically commanding higher prices and yielding better views. For example, someone who wants to live on the ocean and may not be able to afford the several million-dollar price tags for a home may opt instead for a condo at a lower price. Even so, some condos run into the millions of dollars depending on location and features.

Condominiums are communities unto themselves. The beauty of condo living is that most of the upkeep of a regular single family home is eliminated. There is no lawn to cut, no shrubs to trim, and you won't ever be asked to paint the building in your spare time on the weekend. Amenities range from the bare bones with a swimming pool and fitness room, to total luxury with full-time concierge, doorman and valet, room service, spas, and restaurants.

Q & A WITH MARK ZILBERT

I couldn't think of a better way to introduce you to the condo lifestyle than with a question and answer session with one of South Florida's top condo REALTORS®.

Mark Zilbert is a South Florida REALTOR® specializing in the luxury condo market. He has been featured on NBC's Today Show, ABC's 20/20, and has been quoted in numerous newspapers and magazines as the expert in his market. Mark was kind enough to answer some questions about his baby boomer clients, the condominium lifestyle, and the real estate market in general.

Ryan: What percentage of your business would you attribute to baby boomers?

Mark: I would estimate that about 25 percent of my business is selling real estate to baby boomers. What is interesting is that many of my baby boomers are purchasing properties now, but won't retire for another 10-20 years.

Ryan: Do you see that number growing?

Mark: The number is growing significantly. The aging US population is showing a trend of moving to warmer climates. On the East Coast we see over 1000 people moving to Florida each day. Many of these are baby boomers. I predict dramatic growth in these numbers in the next 10 years.

Ryan: What about condo living appeals to them?

Mark: Condo living allows a baby boomer to continue the lifestyle to which they have become accustomed to (at all levels of the economical scale), yet the burden of home maintenance is mostly removed. Also, most new condos offer high levels of services, enabling a much easier lifestyle.

Ryan: What amenities and services are they most interested in?

Mark: Baby boomers have two main criteria: ease of lifestyle and security. The types of services that make day-to-day living easier are at the top of the list. Boomers want 24-hour staffing, including valets for parking and groceries, plus they want the comfort that a condo property is manned 24 hours a day (with technology assistance as well) for their peace of mind. In higher end condos, boomers want spa and pool services (with attendants).

Ryan: What advice would you give to a boomer considering Florida for a second home or retirement?

Mark: Despite what some of the press has been writing about "oversupply," we cannot lose sight of the fact that we are running out of land and the population is exploding. In 2009 it is estimated that Florida will have more residents than New York State. My advice to boomers is to buy property now, and not down the road when prices are once again expected to soar. Many boomers are buying properties today as second homes, but will ultimately pay off the mortgage and retire in them.

For more information on condos in South Florida, visit Mark Zilbert's website at http://www.zilbert.com

TOWNHOUSES

Townhouses can be considered sort of a happy medium between a single family home and a condo. Townhouses are two-story structures that are similar to single family homes in that they sit on their own piece of land.

They are also like a condominium in that they are attached to one or more other homes. They commonly include either a one or two-car garage and also a front or back patio for lounging outside.

The outside of the home is typically taken care of for you, you don't have anyone living directly above you, and there is frequently a small piece of the yard for you to call your own in which you can plant annuals or a rose bush, etc. (often subject to community restrictions). These benefits account for the rise in popularity of townhouse living in Florida.

In most communities, townhouse owners are assessed for the maintenance of the common areas (parts of the community owned equally by the home owners), as well as any amenities provided such as swimming pools, tennis courts, and pavilions.

These assessments can occur monthly, bi-monthly, quarterly or yearly. Most likely these fees will not be figured into your mortgage, so you will have to make a separate payment when it is due. Again, you should review the budget and the association rules before you make a purchase.

ATTACHED "VILLAS"

Available in some communities, attached villas are very similar to townhouses in that they are attached to another unit. However, they are only one story and therefore have no stairs, which some people find to be an attractive feature.

MANUFACTURED HOMES

Close your eyes and step into a modern manufactured home. Now open them. Are you sure that you're really in a manufactured home? You see drywall, crown molding, tile, hardwood floors, a fireplace, decorative niches, and archways. Then look at the floorplan and layout, it seems that this can't be a manufactured home!

Manufactured homes have come a long way from the long and narrow tin cans on wheels of the 50s, 60s, and 70s and have evolved into a logical, economical, and safe choice for many would-be homeowners. Affordability is one of the main factors driving the increase in manufactured home ownership. Manufactured homes cost considerably less than their site-built counterparts, sometimes 25-50 percent less, in fact.

Money is one thing you say, but are they safe? Today's manufactured homes are built in quality and environmentally controlled factories and adhere to current federal building codes. This, combined with the fact that they are anchored to the foundation on which they sit, a process that is overseen by local building inspectors, means a safe and secure home that can withstand the elements. Manufactured home builders' websites are often filled with testimonials of how their homes have been able to withstand hurricane force winds just as well and sometimes better than some site-built homes.

Before you jump in though, some caveats to consider: Though they may be built to withstand winds over 100 mph and are up to federal codes, manufactured homes are still feared by many insurers.

It can be tough to find insurance on your manufactured home at a reasonable rate. Also, when hurricanes threaten Florida, especially near the coast, manufactured home communities are almost always under mandatory evacuation orders, even if site-built home communities surrounding them are only under voluntary evacuation orders. That might be something to think about if you don't want to have to pick up and go every time the wind blows.

MANUFACTURED HOME COMMUNITIES

Most baby boomers entertaining the purchase of a manufactured home in Florida will be considering manufactured home communities that offer a full array of amenities like golf, tennis, swimming pools, clubhouses, and restaurants and bars. It's not just the home, it's the lifestyle that boomers are after, and many developers have realized this and are offering it to the manufactured home buyer.

However, in many (though not all) manufactured home communities in Florida, you do not own the land your home sits on, the developer does. One of the main factors in a home's ability to appreciate is its location and land value, something that in this arrangement you have almost no stake in. This is often times a thorn in the side of residents, but it is what it is. If you really want to live there, it's simply something you'll have to deal with.

Also, the developer will pay the taxes and provide the services outlined in the developer agreement such as grounds maintenance, lawn care, security, and the like, and in turn will charge you a fee, commonly referred to as "lot rent." This is a source of revenue for the developer. The developer is providing you certain services, and you are paying him for providing them. Likewise, when he has an increase in costs or taxes, these increases will be passed on to the homeowners.

One of the best ways to find out more about what owning and living in a manufactured home in Florida might be like is to talk with people who live in a manufactured home. If you are curious, when visiting an area spend some time driving around a manufactured home community and talk with some residents if possible. Most will be glad to share their experiences with you, whether they are good or bad.

For more information on manufactured homes in Florida, visit the Florida Manufactured Housing Association at http://www.fmha.org

5.
TYPES OF COMMUNITIES IN FLORIDA

COUNTRY CLUB

Florida has more golf courses than any other state. There are more than 1,500 golf courses in Florida and most cities have several golf course communities, also referred to as country club communities. Courses can range from fairly modest to extremely upscale. Florida even has a license plate proclaiming it as the "Golf Capital of the World."

Many communities have more than one golf course. Most have at least one clubhouse with such amenities as a fitness center, practice facilities, pro shop, restaurants and bars, banquet facilities, even full service spas, so that you can enjoy a massage after that tough round of golf.

Some golf courses are private, meaning you must be a member or the guest of a member to play there. Membership rates vary among country clubs depending on the location and caliber of the course. Keep in mind that most private courses have a food and beverage minimum, meaning that you have to spend at least "x" amount of dollars in their restaurants and bars within a designated period of time. Thankfully, sometimes purchases in the pro shop can be applied towards meeting your food and beverage minimum. If you lose as many golf balls as I do, you should have no problem reaching your food and beverage minimum.

Many country club communities have equity memberships, which pass from one party to another through the sale of real estate in that community. If this is the case with the home you intend to purchase, be sure that the real estate contract includes the right to the membership. Your real estate agent can help you with this.

Some communities have both a private course and a public course. You can own a home in a community such as this, not be a member and instead choose to play the public course exclusively. Surely, though, if your budget allows you will probably want to be a member of the private course to give your golf game some variety.

Country club communities with a golf course that is always open to the public are also an option. Be aware, however, that public courses tend to be more crowded than private courses, although this can depend on the time of year, the level of the course, and the price you have to pay to play. Some new communities allow the public to use their golf courses until there are enough residents and consequently enough members in the community. This is both good common sense and sound economics.

If you do not play golf, you may want to think twice about buying a home in a golf course community. Many boomers who do not play golf resent the fact that they are sometimes required to help fund its operations through their homeowners' association dues. Whether or not this occurs depends on how the homeowners' association and club budgets are set up, so you might want to look into that before you buy.

> For more information on golf courses in Florida, visit the state's official golf resource at http://www.playfla.com. http://www.floridagolferguide.com is a comprehensive directory to all the courses throughout Florida.

ACTIVE ADULT / 55-PLUS COMMUNITIES

55-plus communities are communities where the majority of the homeowners are over the age of 55. For a community to qualify for the 55-plus designation and to be marketed as such, at least 80 percent of the units have to be occupied by at least one person over 55. A common misconception is that everyone must be over 55 but that simply isn't true. On the other hand, this does not mean that someone under 55 must be allowed to purchase a home. A community—through its deed restrictions—can legally deny someone the ability to purchase a home if they are not yet 55 years old.

Some 55-plus communities have limits on how long relatives such as kids or grandkids can visit, but those instances are usually limited to mobile home parks. While there are still many 55-plus communities in Florida, research suggests that some boomers do not want to move into a 55-plus community due to some of the restrictions involved and because they associate it with being "old." If you fall into this category, you may want to seriously consider a maintenance-free lifestyle community as an alternative. But even still, the business of developing and building 55-plus communities is... excuse me... booming, and those who do it right are experiencing amazing successes.

For a list of 55-plus communities in Florida visit http://fchr.state.fl.us/55+_registered_list.htm

Wiregrass Country Club by Del Webb

One company that focuses on the 55-plus demographic is Del Webb, a division of nationally known Pulte Homes. Pulte's Del Webb division has developed several 55-plus communities around the country with more on the way, and they are considered the leader in innovative 55-plus communities.

One such community is called Wiregrass Country Club in Wesley Chapel, Florida, which is near Tampa. Wiregrass will feature over 2,000 homes ranging in size from 1,100 to 2,300 square feet, designed specifically—like all Del Webb homes—for the needs of the over 55 market.

Located close to gulf coast beaches and within easy driving distance of Tampa, St. Petersburg, and Clearwater, Wiregrass will provide all the social interactions that most of the 55-plus market desires with a clubhouse, indoor and outdoor pools, golf courses, and fitness centers.

For more information on Del Webb homes and communities visit http://www.delwebb.com

MAINTENANCE-FREE LIFESTYLE COMMUNITIES

If cutting grass, landscaping, painting, pressure washing, and general upkeep of the exterior of your home are appealing to you, skip to the next section. Still with me? Okay then, a maintenance-free community might be for you. While some maintenance-free communities are designated 55 and better, most are not. But because maintenance-free lifestyle communities often have restrictions such as no fences, no swing sets, and no basketball hoops, they tend to discourage many families with young children from moving in.

So in maintenance-free lifestyle communities you might enjoy a little more peace and quiet but at the same time be free to have your children or grandchildren visit how often and how long you like.

Maintenance-free communities are those in which you pay a monthly, quarterly, or yearly fee (sorry, the "free" in "maintenance-free" doesn't refer to the cost) to a homeowners' association or resident association, and in return, the association contracts with outside vendors to take care of certain maintenance and upkeep. Some homeowners' associations fees just include the cutting of your grass and leave the homeowner to take care of other items or contract with vendors directly to have them done. Others include complete landscaping such as shrub trimming, mulching, fertilizing and spraying of the yards, painting, and pressure washing.

Most maintenance-free communities are highly amenticised, with clubhouses, swimming pools, billiard and card tables, craft rooms, fitness centers, and activity directors. The idea is that you fill your time doing the things you enjoy, while leaving the work to someone else.

The Palms at New Smyrna Beach

Maintenance-free communities are gaining in popularity all over Florida. A new maintenance-free community located in New Smyrna Beach, just south of Daytona Beach and about 30 minutes from Orlando, is called The Palms at New Smyrna Beach. Though still in early planning and development stages at the time of this writing, The Palms is the second maintenance-free community built and developed by South Daytona based Winston-James Development, Inc. The Palms is a follow-up to their first maintenance-free community, Villages of Royal Palm, located in Port Orange, which at the time of publication has homes still available.

The centerpiece of The Palms will be the planned 20,000-plus square foot clubhouse which will be full of amenities such as a complete fitness center, aerobics room, library, meeting space, ballroom, catering kitchen and bar, and an equally impressive outdoor recreation area with a beach entry pool, lazy river, fire pits for ambiance at nighttime gatherings, an outdoor cabana/snack bar, and other impressive features. Best of all is the fact that someone else is taking care of cutting your grass and trimming your shrubs at your house, leaving you plenty of time to enjoy all the community has to offer.

For more information about The Palms or the Villages of Royal Palm visit http://www.winston-james.com or http://www.villagesofroyalpalm.com

DISCLOSURE: The author has been an independent consultant for Villages of Royal Palm and also owns a home there.

RESORT AND CLUB

Imagine coming to Florida and arriving at your condo, villa, or home and all your favorite groceries are in the fridge, your linens are freshly cleaned and beds made, the wine is chilling, and while you've been away you have actually been making money by letting the management rent out your home while you've been gone. This scenario describes what it might be like to own a home in a resort and club community. In these types of communities, which are often amenticised just like a five star resort, you can elect to have your home in the rental pool, or keep it out, whichever you prefer.

Resort and club communities are great for boomers who might not be quite ready to make a full-time move to Florida, but would prefer to take baby-steps in that direction. However, with residents in and out all the time, it can be hard to forge solid relationships with your neighbors, a factor that may make this option less attractive for some.

Reunion Resort and Club

An example of a Resort and Club community is Reunion Resort and Club, located in Orlando. Reunion Resort and Club is being developed by the Ginn Company. The Ginn Company is a developer that you will definitely be hearing more about in the near future. They have several projects under development in Florida, the Carolinas, and the Bahamas. They recently began sponsoring and hosting tournaments on the LPGA Tour and company founder Bobby Ginn recently bought a major stake in a NASCAR race team.

Encompassing over 2,000 acres within a gated community setting, Reunion Resort and Club has everything a baby boomer could want in a resort community. Reunion has 36 holes of championship golf with more on the way, a unique pool pavilion with a water park and kids' area, a multimillion dollar clubhouse with fine dining, a fitness center, tennis center, spa, stables with horseback riding, and more. For boomers who like knowing their fridge will be stocked with your favorite items upon arrival, concierge service is available.

Reunion Resort and Club offers homes, townhouses, and condos from the $600's to several million. As a property owner you can choose to have your home rented out to resort guests for a fee that you will split with the developer. If income from your property is of no concern to you, you can have your property excluded from the rental pool. Or if you are looking for a second home initially, you could buy your property, use it when you want, have it make you a little bit of money when you're not there and when you are ready to live there full time, just take it out of the rental pool.

> For more information on Reunion Resort and Club visit http://www.reunionresort.com. Or, for more information on other projects the Ginn Company has in development visit http://www.ginncompany.com

GATED COMMUNITIES

Just as in several other states, maybe even your own, gated communities are located all over Florida. Gated communities are gaining in popularity across the nation, especially in the Sunbelt. They can either be manned, with guards posted at the gates and patrolling the streets regularly, or they can be unmanned, with arms or gates that open when you press a button on your garage door opener or enter your secret code in a call box. Guests will either be required to stop and speak to the guard or call your home from the call box before proceeding into the community. While this can sometimes be inconvenient for some people, there is no doubt that gated communities do a good job at keeping solicitors, sightseers, and general riffraff out of the neighborhood, as well as protecting and enhancing the value of the homes in a community.

If your home in Florida is just going to be a part time residence, you might enjoy the added peace of mind that a gated community can give you while you are away. Guards in some communities will even check your doors and windows for you while you are gone. Some can act as a sort of concierge service, accepting packages for you and putting them aside for you until you return. When considering a gated community, be sure to ask your salesperson or real estate agent what level of service you can expect from the guards in the community you are considering.

If you get a chance, speak to a guard and see if they can give you any tips either on the community or the area you are considering. Guards typically see hundreds of people every day and therefore have their fingers on the pulse of the community.

Do not let the fact that a community is gated lull you into a false sense of security. No community, gated or otherwise, is immune to crime. Crime can happen anywhere, it does not discriminate based on zip code. Remember to keep your doors locked, garage door closed, and store any valuables in a safe place.

6.
HOMEOWNERS' AND CONDOMINIUM ASSOCIATIONS

The general idea behind a homeowners' association (HOA) is that you have a group of people elected by the residents who make up the board directing the homeowners association. The main duties are to 1) represent the best interests of the residents of the community especially in the capacity of protecting home values through the implementation and enforcement of rules, known as covenants and restrictions and 2) to assess and collect homeowners' fees to help pay for the upkeep of common areas of the community as well as any other areas provided for in the covenants and deed recordings.

FEES AND DUES

Homeowners' association dues vary widely depending on the amount of amenities that are provided to the homeowners. Some just cover the maintenance of the common areas including medians, right of ways, lakes, and ponds. Other dues can cover things such as upkeep of the streets (if they are private streets), and streetlights. Some communities negotiate for a group rate on cable TV or Internet access with service providers. You may be charged fees for those services monthly, quarterly, or yearly. Failure to pay your homeowner's dues can result in the association placing a lien on your property and eventually foreclosing if you get far enough behind on your payments.

As a prospective purchaser in a community, you are entitled to and encouraged to review the budget. When deciding whether a homeowners' association's dues are a good deal or not, add up what you think it would cost you to obtain the services provided on your own. Don't forget the aggravation the association saves you by not having to deal with finding and scheduling the services and vendors yourself.

If you are buying a home in a new subdivision where homes are still under construction, odds are that the developer still controls the homeowners' association. Until control of the HOA is given to the resident owners, called turnover, which the state of Florida requires to occur when 90 percent of the units in a community have sold and closed, the developer is still responsible for maintaining the public aspects of the community (streets, common areas, etc.) and carrying out the duties of creating a budget for the Association and setting HOA dues accordingly. Oftentimes the developer will over-subsidize the budget, in order to keep the initial HOA fees low, in an effort to attract more buyers. But when turnover occurs, and the developer is no longer subsidizing the budget, homeowners can be hit with a sharp increase in their HOA dues. Before purchasing in a community where the developer controls the HOA, make sure that you carefully review the budget to make sure everyone is paying their fair share, or if that is not the case, try to reasonably figure out what your dues might be when control of the development turns over.

UNDERSTANDING THE HOA BUDGET

On the next page I have provided a sample budget for an imaginary community. I made up the figures, so don't spend too much time wondering how I came up with some of them. The point was to simply let you see what a proposed budget might look like when you are looking at a community with an association, be it a homeowners' association, or a condo association. A budget that you are given could have more expense categories or it could have less, depending on the level of services provided. Regardless of the budget format, there are three things that you will want to look for whenever an actual budget is placed in front of you.

First, look at the Developer's Contribution. In the case of the sample budget provided, it is $105,000 in 2007. It's important to know how much the developer is kicking in to keep the HOA in the

BOOMER PARADISE HOA: Proposed Budget
January 1, 2007 through December 31, 2007

	2007 Proposed	Buildout (Completion)
INCOME (# of Homes)	150	200
Dues (@ $200/month)	360,000	480,000
Late fees	0	0
Initial Fees	0	0
Developer Contribution	105,000	0
TOTAL INCOME	465,000	480,000
EXPENSES		
Homes		
Lawn Care	55,000	62,000
Spray and Fertilizing	12,000	15,000
Pressure Washing	8,000	11,000
Cable	60,000	70,000
TOTAL HOMES EXPENSE	135,000	158,000
Grounds		
Labor	30,000	40,000
General Maintenance	15,000	20,000
Grounds Maintenance	65,000	75,000
Spray and Fertilizing		
Common Areas	7,000	9,000
Water & Irrigation	8,000	10,000
Electric Power	23,000	25,000
Lake Maintenance	15,000	15,000
Gate Maintenance/Repair	8,000	8,000
TOTAL GROUNDS	171,000	202,000
Clubhouse/Community Center		
Lawn/Landscaping	25,000	25,000
Pool Service	30,000	30,000
Cleaning	6,000	7,000
Electricity	20,000	20,000
Water and Sewer	5,000	5,000
Irrigation	1,000	1,000
Repairs/Maintenance	3,000	3,000
Cable/Internet	1,500	1,500
Phone	500	500
Activities Director	30,500	30,500
Legal Expenses	1,500	1,500
Insurance Liability	10,000	10,000
Insurance Building	10,000	10,000
TOTAL CLUBHOUSE	144,000	145,000
Reserves for Replacements	15,000	15,000
TOTAL EXPENSES		
AND RESERVES	$465,000	$520,000
Profit/Loss	$0	($40,000)

black. By doing some quick and dirty math, you can determine how much your dues might increase after the HOA turns over to the residents. In this case, simply take $105,000 divided by 150 (current number of homes) and you are left with a $700 budget shortfall for each home. Divide $700 by 12 to get the monthly amount of $58.33. In theory (keep in mind these are all imaginary numbers), your dues would be $58.33 higher per month, post turnover than they actually are right now.

The next item to look at is the projected Loss at Buildout. In the example, this figure is $40,000. You can also use this figure to determine how your dues will be affected. Take $40,000 divided by total number of homes (200) and you come up with a $200 budget shortfall. This amounts to $16.66 extra per month that each homeowner would have to pay.

Last, but not least, take note of the "Reserves for Replacements," In the sample budget, these reserves are grossly under funded. A good reserve fund should account for at least 10 to 20 percent of an association's annual operating budget. If it takes $480,000 to run each year, then the reserve fund should be at least $48,000. The reserve for replacements fund is there to pay for things like the repair of streets (if they are private, usually in a gated community), a new roof for the clubhouse, the resurfacing of the swimming pool, and other items that will come up after a few years of wear and tear. It is very important that it be well funded and maintained.

COMMON RULES AND REGULATIONS

Another aspect of communities with homeowners' associations is that most involve rules and regulations, or covenants and restrictions (C and Rs) also referred to as covenants, conditions, and restrictions (C, C and Rs). Be sure to ask for a copy before you sign any purchase agreement, and make sure that the agreement is contingent on (depends upon) your understanding and approval of the covenants and restrictions and rules and regulations.

Some common rules and regulations that may be included in the documents are rules regarding:

Fences

Some communities have restrictions on what type of fence you may have, the material it can be made of, how high it can be, or if any fences are allowed at all. If a community you are considering does not allow fences at all, and you have pets that require being fenced in, you might have to consider an invisible fence.

Playground or sports equipment

Basketball hoops are not allowed in more and more communities, while some allow portable basketball hoops as long as they are stored in the garage when not in use. Swing sets and slides are also commonly not allowed because of how they can deteriorate in appearance, and in maintenance-free communities where lawn care is included they are a hindrance to the easy cutting of your lawn.

Parking

Overnight or long-term street parking are often not allowed. This is as much a fire and police safety issue as it is an aesthetic issue. Boats and trailers are usually not allowed to be stored outside, so you must find room in your garage or park them offsite.

Changes to the exterior of your home

Most homeowners' associations require that an architectural or design review committee approve any changes you wish to make to the exterior of your home. This includes things such as adding a screened-in patio, swimming pool, or painting your home a different color. Even changes to your landscaping must sometimes be approved.

There is usually a form they have you fill out on which you must describe in detail any changes you plan to make, including a list of materials to be used, who will do the work, and so on. You are also typically required to submit any drawings or plans that show how the change will look when complete. This is to keep everything in the neighborhood looking nice and congruent.

Pets

Some communities have restrictions on the number of pets you may have in a home, as well as the size. These are typically implemented to reduce the number of potentially aggressive dogs such as pit bulls, and are most common in condominiums or townhouses due to the close proximity of your neighbors. Also, most communities and municipalities now have rules requiring you to pick up after your pets. Be mindful of these rules and laws, especially if the area you are moving from had no such ordinances, as you can be heavily fined for ignoring them.

PROTECTION OF HOME VALUES

It can sound like a pain to have to pay these fees and abide by these restrictions especially if you are coming from a community that doesn't have any fees or restrictions. But all these fees and rules, as inconvenient as they may sometimes seem, do serve the important purpose of protecting your home values. If you are going to pay a quarter of a million dollars or more for your new home here in Florida, you want to know that someone is looking out for you and your investment. Ask any reputable real estate agent or property appraiser and they will tell you that communities governed by homeowners' associations have the best track record of preserving and increasing home values.

DECIDING IF AN HOA IS FOR YOU

So, based on the above information, do you think a community with a homeowners' association is for you? If you're at all like me, the answer is a resounding yes. I like knowing that my best interests are being looked after and my home value is being protected. You basically just have to weigh out the pros and cons of living in such a structured environment. While it's not for everybody, I think that most people, especially boomers such as you will ultimately choose to live in and be happy in communities with a homeowners' association. I think it's best for your lifestyle and the future value of your property.

CONDO ASSOCIATION FEES

As an owner of a condominium you will be responsible for paying condo fees. Before buying a condo, make sure these fees have been explained to you in writing. You should also ask to see the budget. When buying a resale condo in Florida you have a three-day "cooling off" period (7 days for new condo construction) during which you may ask to cancel your contract. This is so that potential condo buyers have ample opportunity to examine and understand the condo fees, rules, and budget. Remember though that this only applies to condos, the same "cooling off" period does not apply to any other type of property.

The condo fees are collected to pay for things like maintenance of the exterior of the condo, including insurance on the building, maintenance of the common areas, such as the grounds, swimming pool, and other amenities. Quite frequently in a condo the condo fee includes water, sewer, and garbage service. This is often more convenient for you: almost no one complains about having a few less checks to write.

SPECIAL ASSESSMENTS

Eventually, if you live in a condo (or even in a homeowners' association) long enough, you may fall prey to what is called a special assessment. A special assessment is sometimes a necessary evil, and is used to pay for items such as a new roof or unexpected repairs beyond ordinary maintenance. Your condo's budget should have a reserve set aside for unexpected events, but sometimes if there is not enough money to pay for what needs to be done, unit owners will be assessed. If you are on a shoestring budget or have a fixed income with little reserves, you may want to rethink a condo because just one special assessment can put you in the red.

Also note that failure to pay any of your condo fees or special assessments can result in the condo association placing a lien on your property, which can eventually lead to foreclosure.

> For more information on your rights and obligations as a condo owner visit: http://www.condo-laws.com

COMMUNITY DEVELOPMENT DISTRICTS

Not to be confused with a homeowners' or condo association, a community development district (CDD) is a means used by local governments and developers to shift the burden of developing infrastructure, maintaining roads and landscaping, building clubhouses and other improvements to the homeowners in that district. The way CDDs work is the CDD, run by a board that is chosen by the developer, issues bonds to pay for the infrastructure and other community improvements that new homeowners have to pay back over the course of a number of years, usually 20 to 30. The amount homeowners are assessed for this is added to their tax bill. This can be an unexpected extra expense if you are not familiar with the rules of the CDD.

A lot of people think that as more people move into the community, the amount of assessment will go down. This is not always the case. CDDs can be used to fund clubhouses: if the clubhouse is running a growing deficit, who do you think pays for that? Of course, the homeowner. CDDs are not necessarily a bad thing though, as they can provide a community with amenities and services it otherwise might not get. You just have to be careful and know what you are getting yourself into before buying in a CDD. There are new CDDs popping up all the time so always be sure to ask when buying a home if it is in a CDD.

You can learn more about Florida CDDs and search a list of current CDDs by county at: http://www.floridaspecialdistricts.org/OfficialList/criteria.asp

7.
PROPERTY TAXES

When people who are looking to buy in Florida ask me about the taxes I like to share a little bit of humor. I say Florida has no state income tax, so they make up for it with real estate taxes and speeding tickets. Speeding tickets is the funny part. The real estate taxes, on the other hand, are not so funny for some. The real estate taxes that you pay on a home can vary widely depending on what city and what county you are buying the home in. For example taxes in 2005 on a $250,000 home in Dade County (Miami area) could be $5188.29, and taxes on the same priced home in Sarasota County could be $3412.59.

Why the difference? Well, several factors are at work here, but the main thing is that the taxes will usually be higher in areas that are experiencing rapid population and housing growth. When rapid growth happens some local governments cannot provide the level of services expected of them without raising taxes. This usually happens because city governments didn't anticipate the rapid growth and must then play catch-up. Had they foreseen the growth, it might be a different story. They could have used the expanding tax base from more people moving into the area to increase the amount and level of services that would be needed such as building new roads and infrastructure, providing adequate schools, police, medical, and fire services, and hiring more public servants to oversee and run them.

PROPERTY APPRAISAL

The property appraiser's office has the task of putting a value on your home. This will help determine the amount of tax you will be required to pay. The property appraiser is not, however, the person who determines what your taxes will be. The local government does that when they set the millage rate.

Luckily, most of the time, you won't pay taxes on the entire price of your home. In Florida, property appraisers have a duty to assess your home at "just value." The typical property valuation is targeted between about 85-95 percent (but these are sometimes lower and sometimes higher) of what they think a particular property would sell for. If you just purchased a property, you are assessed at 85-95 percent of the amount you paid for it, that is, your contract price.

The property appraiser's job involves figuring out a reasonable range of values that buyers would pay for a particular property. Property assessments are usually set at the lower end of that range, which is normally around 85 percent. This is a practice used in almost all Florida counties. You will have to check with the property appraiser's office in the area you are considering to determine where in this range they prefer to target.

Many people wonder why the figures of 85-95 percent are used and not 100 percent. The lower figures are used to allow for closing costs, transfer taxes, and real estate commissions that may have been built into the final sales price but are not really part of the "value" in the home.

Some property appraisers are even taking into account the recent run-up of home prices in 2004 and 2005. Most realize it was a high percentage of investor demand that was driving the increase in values, as opposed to "real" demand, and to compensate they are assessing some homes in the 70 to 85 percent range.

While the above information is good to know, in order to get the possible "worst-case scenario" idea of what your taxes will be, use the 100 percent value for your financial planning. Then when you get your tax bill, if it happens to be lower, you will hopefully be pleasantly surprised.

MILLAGE RATE

An essential element to figuring out how much your taxes will be is the millage rate, commonly referred to as "mil rate." The millage rate is expressed as "mils per thousand." For example if the millage rate is "22.55", then you will pay $22.55 per $1,000 of assessed value.

Each taxing district will set its own millage rate which can be determined by dividing the total proposed budget of the taxing district (city, county, school district, etc.) by the total taxable value of all real estate in the district after exemptions are deducted for.

You will likely be taxed by your city, county, school district, water management district if there is one, and others. It's important to get a whole tax picture view before deciding on an area. The local property appraiser's office will usually be your best resource for this.

It is also important to note that real estate taxes in Florida are paid in arrears, and you will have an opportunity to get a small discount for paying them early.

For links to the property appraiser's offices throughout Florida visit http://myflorida.com/dor/property/appraisers.html

How to figure out what your taxes (non-homesteaded) will be on a home you are purchasing:

Price you pay for the home (Contract price)	$250,000
times	x
90% (assuming middle of 85-95% range)	90%
equals	=
"Just Value"	$225,000
Divided by 1,000	/1000
equals	= 225
Times mil rate (we'll assume 20 mils per thousand)	x 20
Equals your tax liability	$4,500

STATE OF FLORIDA HOMESTEAD EXEMPTIONS

The state of Florida does provide some much needed tax relief in the form of homestead exemptions.

$25,000 Homestead Exemption

Florida's constitution provides homeowners the right to receive a homestead exemption provided they meet certain residency requirements. Every person who has "legal or equitable title (you own it) to real property (your home) in the State of Florida and who resides on the property on January 1, and in good faith makes it his or her permanent home is eligible for a homestead exemption." This exemption reduces your "just value" by $25,000, which can save you around $500 on your property tax bill, depending on what the millage rate is for your area.

You have to make application for the exemption between January 1 of the previous year and March 1 of the year you want the exemption. For example, for the 2007 tax year, you would have been able to apply for homestead exemption from January 1, 2006 to March 1, 2007.

When filing your homestead exemption for the first time, you will be asked to provide evidence that you are a legal resident, such as a voter registration card or a Florida driver's license. Most counties have automatic renewal programs so there is no need to reapply each year as long as you are in the same home. If you move, however, you will need to reapply.

If you are a part-time resident using your home in Florida as a vacation home or second home, you will not be eligible for this exemption.

OTHER POSSIBLE EXEMPTIONS

In addition to the standard $25,000 Homestead Exemption, there are other exemptions available that could possibly reduce your tax bill.

Additional Homestead Exemption for Persons 65 and Older of up to $25,000

This exemption is available on a county-by-county basis, and not all counties offer it.

$500 Widow's and Widower's Exemption

$500 Disability Exemption

Any Florida resident who is totally and permanently disabled may claim this exemption with proper documentation of the disability.

$5,000 Disability Exemption for Ex-Service Members

An ex-service member who is disabled at least 10 percent in war or another service-related incident might be entitled to this exemption.

$500 Exemption for Blind Persons

In order to claim this exemption, you must have a certificate of blindness issued by the Division of Blind Services of the Department of Education, the Federal Social Security Administration, or the Veteran's Administration.

These exemptions can be combined with each other. Certain other rules and restrictions apply to the above exemptions. You should contact the local property appraiser's office for more information on the county in which you are interested. For links to the property appraiser's offices throughout Florida visit: http://myflorida.com/dor/property/appraisers.html

SAVE OUR HOMES

The Save Our Homes amendment is a piece of legislation, loathed by many and loved by some, which helps limit the tax burden of those residents who own homesteaded property in Florida. Over the past five years home prices rose by as much as 100 percent or more in many parts of Florida. Most people would have a hard time handling a property tax bill increase of 100 percent in such a short period of time. In fact, in such situations, many homeowners of modest means on fixed incomes would no longer be able to afford the home they live in because they wouldn't be able to pay the taxes.

The amendment states that the assessed value of a homesteaded property cannot exceed 3 percent of the assessed value of the property for the prior year or the percentage change of the Consumer Price Index as reported by the U.S. Department of Labor, Bureau of Labor Statistics, whichever is lower.

For example, the Consumer Price Index rose 3.40 percent in 2006, so the increase in assessed value was capped at 3.00 percent. Let's assume that between 2005 and 2006 the value of your home in Florida doubled from $100,000 to $200,000. (This is not likely to happen in such a short period, but it makes for a clear example.) With the Save Our Homes cap, you could only be taxed on $103,000 ($100K + 3 percent) instead of the full value of $200,000.

PART-TIMERS AND LANDLORDS BEWARE!

As we are all smart enough to know, when taxes are cut or held down for one group, they must be passed on to other groups. Groups that bear the burden of the Save Our Homes legislation are:

- Property owners who reside in Florida only part-time and cannot claim a homestead exemption
- Local business owners as the Save Our Homes cap does not apply to their business real estate
- Tourists who are forced to pay higher taxes at hotels
- Landlords as they also cannot homestead their rental property. The landlord's increasingly higher taxes are then passed on to their tenants in the form of rent increases.

THINKING OF DOWNSIZING? THINK ABOUT THIS FIRST

Homesteaded property owners who wish to downsize often times can't afford to because the taxes on a smaller home that they buy could be much higher than what they are currently paying, because the cap does not transfer with the owner. When the homeowner moves, they lose their Save Our Homes benefit until they have been in their newly homesteaded property for one year. By then it's too late.

When Save Our Homes was passed in 1992, the inequities that it would later cause were hard to have predicted. Not many people could have foreseen the explosion of property values across the state and the problems that a tax cap would cause.

For years several groups adversely affected by this law, including business owners, landlords, and part-time residents that do not qualify for homestead exemptions and subsequently the Save Our Homes cap, have been trying to get the Florida legislature to get rid of or, at least drastically change the Save Our Homes law.

PROPERTY TAX DISCLOSURE

Once a home changes ownership, it is reassessed at its full value, and the Save Our Homes amendment does not go back into effect until a year after the property is homesteaded again.

Consider this scenario. You are looking at a house on the market for $400,000. The Multiple Listing Service printout that your agent gives you on the property states that the amount of property taxes paid by the owner for the prior year was $2,500. It is a very common but costly mistake for buyers to assume their property taxes will be that same amount. I've heard "but it says right there on the sheet, Property Tax: $2,500" too many times. The truth is that your taxes will be much higher because you will be required to pay taxes on the amount that you pay for the property. The current owner is paying taxes based on what they paid for the property, which assuming they bought the property even just a couple of years ago, could be considerably less.

A sharp agent will catch this and make you aware of what your new property tax burden could be. The Florida Legislature thought that not enough people buying homes were being made aware of this fact so they introduced a new Property Tax Disclosure which must be in all residential real estate contracts after January 1, 2005.

It reads:

PROPERTY TAX DISCLOSURE SUMMARY

> *A buyer should not rely on the seller's current property taxes as the amount of property taxes that the buyer might be obligated to pay in the year subsequent to purchase. A change of ownership or property improvements triggers reassessments of the property that could result in higher property taxes. If you have any questions concerning valuation, contact the county property appraiser's office for information.*

This clause is very easy to miss, as it is usually not presented on a separate disclosure, but is included in the standard contract form. Hence, many people do end up missing it and later wishing that they had read this book!

The best resource for calculating what your taxes will be on a particular property is, again, the local Property Appraiser's office. They are there to help you. Here's the website listing for the Property Appraiser's offices in Florida: http://myflorida.com/dor/property/appraisers.html

Some even have calculators on their website that you can use to get an idea of what your taxes will be.

2007 PROPOSED PROPERTY TAX REFORM

Florida Governor Charlie Christ hit the ground running his first month in office in January 2007, announcing his plan to cut taxes in order to "keep Florida's economy vibrant." Among his proposals are:

- Doubling the Homestead Exemption from $25,000 to $50,000.
- Make the Save Our Homes cap portable statewide, meaning that Florida homeowners can take their current tax rate with them when they move.
- Place a cap on property taxes for businesses and landlords of the lesser of three percent or the rate of inflation. This is similar to how the Save Our Homes legislation currently protects homeowners.
- Exempting small businesses from tangible personal property taxes.

There is even discussion of eliminating property taxes altogether and raising the sales tax rate to make up for the lost revenue. Most impacted by these proposals are local governments who would have to trim the fat and possibly a whole lot more from their budgets because of a lower property tax increase each year. Their budget will still grow, just not any faster than three percent per year. For some local governments this may not be enough to keep up with rapid growth.

We will all have to stay tuned to see whether Florida voters approve these proposals in a special election in 2007. Then, it will probably take a couple of years before we are able to determine how exactly these measures are impacting Florida property owners and local governments.

You can keep your eye on major developments of this issue by visiting the "Florida Resources" page at http://www.floridaforboomers.com.

8.
HOMEOWNER'S INSURANCE

With four hurricanes hitting the state in 2004, and the widely publicized destruction caused by Hurricane Katrina the next year, it doesn't take a genius to understand that insurance might be a little expensive or hard to come by in some areas of the state. In fact many news outlets are reporting that the homeowner's insurance industry, particularly in Florida, is in shambles and just a bad hurricane away from being defunct.

In addition to the obvious need to protect your investment, the ability to get homeowner's insurance is of utmost importance when getting a mortgage on your property. No mortgage company will loan you money without you first having insurance on the property. In some instances your mortgage lender can even foreclose on the property if you fail to carry insurance. While no one can predict what the future will bring, I have done my best to compile for you the facts and resources, as they now stand, to help you navigate the homeowner's insurance minefield in Florida.

GETTING STARTED

Presumably you now have homeowner's insurance on your current residence, wherever that might be. My first piece of advice is to ask your current insurance agent if their company writes homeowner's insurance policies in Florida. If you are with a large national insurer with operations in Florida like State Farm, Nationwide, or Allstate, the odds are good that they do, at least at times (more on that in a moment), write homeowner's policies in Florida. By doing this, you are taking the path of least resistance, and you will probably be able to get pretty decent rates through what are called "multi-line" discounts assuming you have other property such as cars, jewelry and the like already insured through them. You are welcome to shop around and price out other insurers, but from what I've seen, if you are comfortable with the company you have now, switching carriers to save a few bucks isn't worth the hassle.

ASK AROUND

If the above scenario does not work out for whatever reason, my next step would be to ask any family or friends presently living in Florida who they have as their homeowner's insurance carrier. If they have no useful information, ask a real estate agent or potential new neighbors for a recommendation.

Because of the ever present risk from hurricanes as well as rising re-insurance rates (yes, insurers get insurance to protect themselves against losses on the insurance they issue you), many insurers in Florida have recently dropped customers in damage prone areas, stopped writing new policies in those same areas, or stopped writing new policies in the state altogether. You are going to have to do some serious legwork, phoning different agencies to see who is doing what at the time when you need insurance. Many have a one policy out, one policy in type of arrangement, where they will place you on a waiting list and when a policyholder does not renew for whatever reason, they can pick you up.

NEXT OPTION

If you've tried all of the above and still are not able to acquire coverage, your next step is to visit the website of the Florida Market Assistance Program at fmap.org. FMAP is a free on-line referral service, created by the Florida legislature and designed to connect those who are not able to find coverage with insurers who are able to write new policies. You register at the website and submit a request for quotes. Agents will typically then call or e-mail you if they can offer assistance to you based on where your home is located, its age, and other factors.

For more information visit http://www.fmap.org

THE LAST RESORT

If all of the above options fail and you are not able to find private homeowner's insurance coverage, Citizens Property Insurance Corporation is your last available option besides self-insuring. This is commonly referred to as "going naked," which I promise you is not as fun as the name suggests.

The Florida Legislature created Citizen's in 2002 to help Floridians who cannot get traditional coverage. More than a million homeowners in Florida turn to Citizens for their homeowner's insurance coverage. Here is fair warning, though. Because Florida law mandates Citizen's rates, they are much higher than the rates of private insurers. This is by design, so that Citizen's does not directly compete with private insurers. So the bottom line is don't turn to Citizens unless you absolutely have to.

For more information visit: http://www.citizensfla.com

2007 PROPERTY INSURANCE LEGISLATION UPDATE

In January 2007, Florida legislators passed a bill that may lower property insurance premiums and bring other insurance relief to millions of property owners. Among the highlights of the 167-page bill are means to lower rates for homeowners by allowing owners to exclude windstorm and contents coverage, allowing owners to choose higher deductibles so that their premiums will be lower, removing the requirement that Citizen's charge the highest premiums, and allowing non-homesteaded properties to be eligible for Citizen's.

In order to further protect policyholders, the bill requires insurance companies to give 100 days notice of their intent to cancel a homeowner's policy that would be effective during hurricane season, requires insurance companies to expedite payments of claims after storms, and prohibits excess profits by insurers.

The bill also makes an effort to expand the property insurance market in Florida by requiring insurance companies that write homeowners policies in other states and auto insurance policies in Florida, to write homeowners policies in Florida.

Insurance companies lobbied heavily against this bill, and there are fears of companies pulling out of Florida altogether. Just like the property tax issue, it's a matter of "wait and see" before we can determine the effects of this legislation.

> You can follow major developments of this issue by visiting the "Florida Resources" page at http://www.floridaforboomers.com.

HOW TO LOWER YOUR WIND PREMIUMS

Despite how gloomy I've made homeowner's insurance in Florida sound, there are ways in which you can save significant amounts of money on the windstorm portion of the insurance premiums you pay each year. Most of these have to do with the manner in which your home is constructed.

For instance, just owning a home that is compliant with the current Florida building codes, (that is, built after 2002, or after 1994 in Dade and Broward counties) can save you nearly 50 percent, depending on the insurance company. All the more incentive to buy or build a new home. If that home is built of concrete block, you save another few percentage points. Also, since the roof—as you might have guessed—is an important component in standing up to a hurricane, certain roof types make your home eligible for reduced rates. Experts agree that a hip roof, or a roof that is sloped on all four sides, will perform the best under extreme winds, and having a hip roof on your home can save you almost 25 percent. Other savings are available for having certain types of protective shutters installed on the windows of your home.

In order to get these discounts, most insurance companies either ask that you get a certification from the builder or they will send some other independent party to your home to verify that the home does in fact comply with the requirements for the discounts.

> For more information on these and additional savings that may be available to you, consult with your insurance agent or visit: http://www.mysafefloridahome.com/insurance.asp

TYPES OF COVERAGE

When shopping for homeowner's insurance, there are various types of coverage available for you to choose from. Perhaps the most important is guaranteed replacement cost coverage.

Guaranteed Replacement Cost Coverage

No matter which insurance company you eventually go with, you need to make sure that all rates that you are being quoted are for what is called guaranteed replacement cost coverage, as opposed to actual cash value coverage. Guaranteed replacement cost coverage means that even if you are insured for, let's say, $200,000, if your home is destroyed and it costs $250,000 to build at today's construction costs to be put back into use as it was before, then that's what the insurance company will pay.

This type of coverage will cost more, as you might imagine, but it provides the policyholder with much more protection. One way to mitigate the rise in your premium is to raise your deductible. When you raise your deductible, or the amount you pay out of pocket to file a claim, your yearly premiums will go down.

Endorsements and Additional Coverage

Your homeowner's insurance policy may not cover certain items in your home against damage or theft. Always have your agent explain anything you don't understand and read your policy carefully.

Items like fine art, coin or gun collections, jewelry and furs, or electronic equipment beyond the standard televisions and DVD players, must usually be accounted for and covered under a separate policy or endorsement for an additional premium amount. But if you have any of these items, it might be wise to look into the coverage available to you.

INVENTORY

Always maintain a current inventory of the items in your home, including pictures or video of your property. This inventory should be kept in a safe place outside of the home like a safe deposit box at a bank. This way, if your home and personal belongings are destroyed, you have evidence of what was damaged to show the insurance adjuster. This step alone can save you several hours, days, or sometimes weeks of hassles and delays in getting your insurance claim processed.

FLOOD INSURANCE

There are two facts that most people do not know. First, flooding is the number one natural disaster in the United States; even properties not near water can be susceptible to flooding. Second, losses due to floods are not covered by your homeowner's insurance policy.

The Federal Emergency Management Association (FEMA) puts out "flood maps" that show which areas tend to be most prone to flooding. I often hear people ask, "is the property in a flood zone?" and usually people describing homes in low-risk areas will say "no, it's not in a flood zone." Well, the correct answer is that every property is in a flood zone. It's just a matter of whether it is in a low, moderate or high-risk flood zone.

Your real estate agent might be able to tell you which flood zone the property you are looking at is in. But it is your insurance agent who will use a Flood Insurance Rate Map or FIRM, to ultimately determine your flood risk. Be aware that federal law requires you to purchase flood insurance if you have a federally backed mortgage and reside in a high-risk area.

FLOOD ZONES EXPLAINED

Here is a break-down of the various flood zones in which you may find a property located.

Moderate to Low Risk Areas (Flood insurance is not required, but recommended)

Zones B, C, and X These are flood zones with a less than 1% chance of flooding each year.

High Risk Areas (Flood Insurance is Mandatory)

Zones A, AE, A1-A30, AH, AO These areas have a 1% chance of annual flooding and a 26% chance of flooding over 30 years.

Zone AR This is a flood zone with an increased temporary risk due to the construction or restoration of a levee or a dam.

Zone A99 Areas with a 1% chance of annual flooding that will be protected by a levee or dam where construction has reached specified legal requirements.

High Risk - Coastal Areas (Flood Insurance is Mandatory)

Zone V, VE, V1-V30 Coastal Areas with a 1% or greater chance of flooding annually and subject to the additional hazard of storm waves. Also has a 26% chance of flooding over 30 years.

Undetermined Risk Areas

Zone D Possible but undetermined flood hazards.

BETTER SAFE THAN SORRY

Everyone, no matter where in Florida they live should carry flood insurance available through the National Flood Insurance Program. It is relatively inexpensive, especially if your home is in a low-risk flood zone. A single-family home is limited to $250,000 coverage for structural damage and $100,000 for contents. I have those amounts of coverage on my home in a low-risk flood zone and my premium is just over $300 per year. For less than a dollar a day, I can sleep easy at night when the rain from the latest tropical system is pouring down. It's a no-brainer.

> For more information on flood zones, flood maps, and flood insurance contact your insurance agent or visit www.floodsmart.gov

9.
CONTRACTS AND DISCLOSURES

Now I will walk you through the standard contract for sale and purchase approved by both the Florida Association of Realtors and the Florida Bar, simply referred to as the "FAR/BAR contract." This by no means should be construed as a complete contract or explanation of it, nor is this legal advice. What I want to do is highlight the major sections of the contract so that when you find your dream home and this contract is placed in front of you, you will hopefully feel a little more comfortable with it.

Please also note that this contract will not always be the contract used. For example, new home builders typically have their own contracts. It is, however, the contract most commonly used in Florida today.

DEMYSTIFYING THE FLORIDA CONTRACT FOR SALE AND PURCHASE

Lines 1 through 15

These have blanks to be filled in with the seller's name(s) and buyer's name(s), the legal description and address of the property, and items of personal property that are included and excluded from the sale. This section also has legal verbiage stating that

CONTRACTS AND DISCLOSURES

the "seller shall sell" and the "buyer shall buy" the property in question.

Lines 16 through 24

This section covers the purchase price, how much deposit is given (including where the deposit is to be held, which is usually a brokerage escrow account, title company, or attorney's escrow or trust account), how much financing the buyer is going for, and how much money will be required at closing.

Lines 25 through 32

Include the time for acceptance of the contract as well as the effective date (the date when both parties came to an agreement and entered into a bilateral agreement). The buyer will typically make an offer, and the seller will have a certain amount of time to agree to the terms or make a counteroffer. If the seller fails to respond in the allotted time frame, the offer can be withdrawn and the deposit returned if the buyer so desires.

Lines 33 through 49

Cover financing. If the buyer is paying cash, there are no contingencies for financing. If you are getting a mortgage to pay for part of the purchase, you should make the contract contingent (check line 35) upon you applying for and receiving this mortgage.

Lines 50 through 55

Include when you will receive the title insurance policy and who will pay for it. Customarily the seller pays, but this is a negotiable point and the buyer is sometimes asked to pay.

Lines 56 through 58

Outline the closing date. This is a date mutually agreed upon and negotiated between the seller and the buyer. Something that comes up from time to time during hurricane season is that an insurance company might not write an insurance policy on a home you are buying if there is a hurricane, tropical depression, or tropical storm within so many miles of Florida. In the case of this, the closing may be legally delayed for up to 5 days after the coverage becomes available again.

Lines 59 through 65

Explain that the seller will provide you with "marketable title" subject to certain restrictions, easements, and limitations. The seller is also stating that there are no violations that would prevent you from using the property for your specific purpose, usually as a residence.

Lines 66 through 69

Discuss occupancy, and basically states that you the buyer, will get occupancy at the time of closing. This can be modified if you agree to let the seller stay in the home past the closing date.

Lines 70 through 73

Goes over provisions typed into or written into the contract. Also, whether or not you may assign the contract is covered here.

Lines 74 through 90

Cover disclosures. This is a very important section that you should be sure to read. It discusses disclosures such as radon, mold, special assessments, lead based paint, homeowners' association disclosures, and more.

Lines 91 through 96

This section covers repair costs and home warranties. Here it is spelled out how much of the repairs deemed necessary by the home inspection report the owner is willing to pay for. This includes figures both for damage caused by termites, and for damage not caused by termites. Also, if you will be receiving a home warranty, it is spelled out here what company it will be with, what the cost will be, and who will pay for it.

Lines 97 through 103

Riders, addenda, and special clauses to the contract are discussed here. It is the best practice to have any special typewritten or handwritten clauses added into the contract to be written by an attorney. However, this does not always happen, and problems can sometimes arise even with the simplest of clauses.

Lines 104 through 112

More disclosures, including a statement indicating that by signing you affirm that you have read the Standards for Real Estate Transactions A through Y which I will go over in the next few pages.

Lines 113 through 123

Spaces for the buyer's signature and seller's signature, as well as blanks for addresses and phone numbers for notification purposes. Also, any listing or cooperating brokers are listed on the bottom.

"FAR/BAR" STANDARDS FOR REAL ESTATE TRANSACTIONS

As previously mentioned, there are 25 "standards" lettered from "A" to "Y" in the standard FAR/BAR contract. Here is a brief explanation of each of them. Once again, you should have an attorney explain to you anything that you do not understand or do not feel comfortable with.

A. Title Insurance

This tells you who must issue your title policy, when you will receive a copy, and what will happen if it is determined that the title is not clear or marketable.

B. Purchase Money Mortgage; Security Agreement to Seller

This only applies in the case of the seller providing you with financing. It tells you what the consequences and timeframes involved are for foreclosure, as well as spelling out the duties of both parties.

C. Survey

This allows for you the buyer, at your expense, to have a state registered surveyor conduct a survey of the property. If you are getting financing, your lender will require this. If you are not getting financing, it is advisable to get a survey done anyway. It is not very expensive and a good thing to have down the road as you will need it if you ever want to get a permit to make changes to your home. The main objective of the survey is to ensure there are no encroachments on the property, which would constitute a title defect.

D. Wood Destroying Organisms

Defines the term "Wood Destroying Organism" and lets you know that you have a right to have a Florida Certified Pest Control Operator inspect the property for their existence. This service is usually piggybacked with the home inspection. Also provides for possible courses of action should evidence of infestation or damage be found.

E. Ingress and Egress

Simply states that you will have access to and from the property for its designated purpose (i.e. to reside there) without infringing on anyone else's property rights.

F. Leases

Spells out what must be done if there is a tenant living in the property you are buying. For example, the seller must give you copies of the lease, tenant names, and any security deposits being held.

G. Liens

Instructs the seller to provide you with an affidavit at closing stating that there are no liens on the property, for example financial liens and construction liens.

H. Place of Closing

This states that the closing should be held in the county where the property is located. The party paying for title insurance usually chooses where the closing will take place, whether it is at an attorney's office, or the office of another closing agent.

I. Time

This section tells you how to account for weekends and holidays. It also includes a very important phrase, "time is of the essence." This means that all dates and times in the contract are of extreme importance and should be followed exactly.

J. Closing Documents

This states which documents must be furnished by the seller (deed, bill of sale, certificate of title, etc.) and which documents must be furnished by the buyer (Mortgage, mortgage note, etc.)

K. Expenses

This standard designates which party shall pay which expenses. Typically, the seller will pay for documentary stamps on the deed and recording, but this is a negotiable point. The buyer is typically expected to pay for their entire loan closing costs as well as the recording of the mortgage.

L. Prorations and Credits

Tells how prorations and credits for homeowner's association dues, insurance, taxes and the like, will be handled.

M. Special Assessment Liens

Any pending special assessment liens will be assumed by you, the buyer, as of the closing, so make sure there are not any pending liens that you are not already aware of.

N. Inspection and Repair

It is very important that you understand this standard of the contract. It says that the seller, unless otherwise specified in the Seller's Disclosure, states that all components of the home are in working order (definition of "working order" is contained within the paragraph). Note that cosmetic flaws (definition of "cosmetic" is also in the paragraph) are not covered here. The buyer is allowed to inspect the home or pay someone with proper credentials, such as a state certified contractor or someone from a firm specializing in home inspections (my suggestion) to inspect the house and report any substandard findings to the seller in writing within 20 days of the effective date of the contract. If you fail to do this, you waive your right to the seller's warranties of unreported defects. The seller is required to repair up to the amount specified in line 93 of the contract, or 1.5% of the contract sales price if this line has been left blank. So, bottom line, get an inspection done, and do it in a timely fashion.

O. Damage Prior to Closing

If the house is damaged by fire or other event before the closing, the seller is required to make repairs, so long as the amount to repair the house does not exceed 1.5% of the contract sales price. If the cost exceeds 1.5% you can take the property as it is and the seller can credit you the 1.5% plus any insurance proceeds at closing, or you may request that your deposit be returned and the contractual obligations of both parties will terminate.

P. Closing Procedure

This standard outlines the procedure that should be followed for closing.

Q. Escrow

This standard describes how funds that are held in escrow should be handled and disbursed.

R. Attorney's Fees; Costs

This states that should litigation occur and legal costs are involved, the winner of those proceedings shall be entitled to recover their attorney's fees and related costs.

S. Failure of Performance

Tells you what the options are of both parties should anyone fail to complete their contractual obligations. Basically, deposit can be returned or kept, depending on the party. Or, either party can sue for specific failure of performance, in an effort to get the other party to go through with the deal.

T. Contract Not Recordable; Persons Bound; Notice; Facsimile

States that the contract is not recordable. All parties and their successors are bound by the contract. Notice given by one party's agent or attorney is as good as that party themselves giving notice. Faxed copies that are legible are considered an original.

U. Conveyance

States how seller should convey title. Also states that any personal property should be covered in a separate bill of sale.

CONTRACTS AND DISCLOSURES

V. Other Agreements

Covers how any other agreements should be handled.

W. Seller Disclosure

Seller states that, to the best of their knowledge, there are no problems with the property, unless otherwise specified, which are not readily observable.

X. Property Maintenance; Access, Repair Standards; Assignment of Contracts and Warranties

States that the seller needs to maintain the property (such as keeping the grass cut and the shrubs trimmed) while the contract is in force. Also states that the seller needs to keep all necessary utilities (power and water) on for buyer's appraisal, inspection, and walk through prior to closing. Any repairs should be made in a good, workmanlike manner using materials equal to the quality of the material present to begin with. Any assignable repair or treatment contracts shall be assigned to buyer at closing.

Y. 1031 Exchange

In the rare case of a party wishing to enter into a like-kind exchange, the other party shall cooperate reasonably to facilitate the exchange. You can read more about 1031-Exchanges in chapter 12.

COMMON FLORIDA REAL ESTATE DISCLOSURES

In addition to the standard "FAR/BAR" contract, there are several other disclosures, addenda, and types of contracts you may see in different situations along the way. Here are a few of the more common ones.

Buying a Home "As-Is"

Sometimes in your search for a home you will come across a home being sold "as-is". This means that you are allowed to have your inspections done, just like a normal sales scenario but the owner will not be responsible for fixing anything you may find wrong with the place. Many people errantly believe that any home being sold "as-is" is a "handyman's special" that will require lots of work to be made habitable. This is simply not always the case.

It's true that is sometimes the owner's reason for selling a place "as-is," but in my experience the reason the seller chooses this route is more often just a matter of convenience. Another reason might be as a point of negotiation. For instance, you may make an offer for twenty thousand less than what the seller is asking for a home, and the seller could agree but add the caveat that you must buy it "as-is." You the buyer, of course, will have to do your own research to see which of these scenarios are true for the house you are considering.

Should you decide to contract for a home being sold "as-is," the contract can be done one of two ways. It can be prepared as a normal contract would be done with an "As-Is" Rider attached to and made part of the FAR/BAR Contract for Sale and Purchase, or it can be prepared as a separate "As-Is" Contract for Sale and Purchase.

"As-Is" Rider to the FAR/BAR Contract for Sale and Purchase

This is a document to be completed and initialed by both parties, buyer and seller, which becomes part of the Contract for Sale and Purchase. This document effectively deletes Standards D (Wood Destroying Organisms), N (Inspection and Repair), and X (as it relates to Repair Standards) of the Contract for Sale and Purchase. The seller is still required to disclose all known facts that affect the value of the property. However, the seller makes no representation as to the physical condition of the property.

As the buyer in this scenario, you are given the right to inspect the property as you would under normal circumstances, but you are also given the right to cancel the contract, should you not be satisfied with the findings of the inspection or do not wish to pay for the repairs that need to be made. You will be given a certain number of days to conduct your inspections and you must let the seller know of your intention to cancel in writing during that allotted time, should you decide you do not want to proceed with the purchase.

"As-Is" Contract for Sale and Purchase

The first thing you'll notice is that at the top of the page "AS-IS" in LARGE BOLD letters, so you can't be confused as to what you are signing. The main differences between this contract, and a normal contract we discussed previously is the deletion of standard "D" (Wood Destroying Organisms) and standard "N" (Inspection and

Repair). Another difference is the addition of another standard, standard "Z". Standard "Z" in the "As-Is" Contract for Sale and Purchase says that you the buyer waive all claims against the seller, or any real estate agent involved in the transaction, should you discover any defects or damage after closing. The seller is still required to disclose all known defects or damage, and of course you have the normal time period for inspections to be conducted.

Homeowners' Association Disclosure

The Homeowners' Association Disclosure, required by Florida Statute 720.401 is one of the more plainly written documents you will come across during your real estate transactions here in Florida. There are eight statements on the current Homeowners' Association Disclosure. Keep in mind this document only applies to communities where there is an obligation to be a member of the Homeowners' Association.

1. Simply lets you know that you will be obligated to be a member of a Homeowners' Association.
2. States that there are or will be restrictive covenants that govern the property.
3. If there is a fee (assessment) to be a member of the homeowners' association, the fee is noted here. Also, if there are any special assessments owed it will be noted here.
4. Gives you warning that if you do not pay the above-mentioned assessments, a lien can be placed on your property.
5. If there is a mandatory fee to use certain facilities in a community, it will be noted here. For instance, some associations have pools or clubhouses that you are required to pay to use.
6. Lets you know whether or not the covenants or rules of the association can be changed without the approval of the association membership. It is preferable that they cannot be changed without the approval of the association membership.
7. Tells you that the statements in the disclosure are only summary in nature and that you should read the covenants and restrictions of the association before purchasing property there.
8. This lets you know that a homeowners' association's covenants and restrictions are a matter of public record and can be obtained from the record office of the county where the property is located.

Just above the signature lines in BOLD CAPS is the following statement:

"IF THE DISCLOSURE SUMMARY REQUIRED BY SECTION 720.401, FLORIDA STATUTES, HAS NOT BEEN PROVIDED TO THE PROSPECTIVE PURCHASER BEFORE EXECUTING THIS CONTRACT FOR SALE, THIS CONTRACT IS VOIDABLE BY BUYER BY DELIVERING TO SELLER OR SELLER'S AGENT OR REPRESENTATIVE WRITTEN NOTICE OF THE BUYER'S INTENTION TO CANCEL WITHIN 3 DAYS AFTER THE RECEIPT OF THE DISCLOSURE SUMMARY OR PRIOR TO CLOSING, WHICHEVER OCCURS FIRST. ANY PURPORTED WAIVER OF THIS VOIDABILITY RIGHT HAS NO EFFECT. BUYER'S RIGHT TO VOID THIS CONTRACT SHALL TERMINATE AT CLOSING."

This simply means that if you completed a contract before receiving this notice, you may be able to void your contract if you so desire, within three days after having finally received this disclosure.

Coastal Construction Control Line Disclosure

If you are fortunate enough to be able to afford property on or near the ocean, depending on which part of the state you buy it (there are areas on the West Coast unaffected by this), you will be asked to sign a Coastal Construction Control Line (CCCL) Disclosure if any part of the property is seaward of the CCCL. The CCCL is a boundary, not a setback mind you, that is designed to protect Florida's sandy beaches and dunes from erosion and to protect their value as natural ecosystems, while still allowing development of private property to continue. Certain restrictions and design specifications exist that you may be subject to if you are planning to build on the ocean.

The Disclosure reads:

> *The property being purchased may be subject to coastal erosion and to federal, state, or local regulations that govern coastal property, including the delineation of the coastal construction control line, rigid coastal protection structures, beach nourishment, and the protection of marine turtles. Additional information can be obtained from the Florida Department of*

Environmental Protection, including whether there are significant erosion conditions associated with the shoreline of the property being purchased.

For more information on the CCCL visit http://www.floridadep.org/beaches

Florida Building Energy Efficiency Rating Disclosure

As the purchaser of a home in Florida you have the right to have the energy efficiency of the home tested. This can be helpful in determining what the annual energy use of the home might be, and how it compares to other homes used as "yardsticks" for means of comparison. You should be notified of this right either before doing a contract or at the time of contract. Most contracts will have verbiage in them alerting you to this fact.

There is a brochure available explaining all that you should know about the Florida Building Energy Efficiency Rating System as well as giving you suggestions to increase your home's energy efficiency at:

http://www.dca.state.fl.us/fbc/committees/energy/EnergyBrochure-110602.pdf

Florida Homeowner's Construction Recovery Fund

The Florida Homeowner's Construction Recovery Fund, previously known as the Construction Industries Recovery Fund, is a fund paid into by all licensed contractors who build, renovate, or remodel homes. The fund, first created in 1993 in the aftermath of Hurricane Andrew, was set up to provide relief to those who suffer monetary damages as a direct result of the actions of a Florida licensed contractor performing one of those functions.

Money for the fund is basically collected as part of a half-cent per square foot surcharge that contractors pay when getting their permits. (I qualify this with "basically" because the half-cent actually goes first to the Professional Regulation Trust Fund, and the Building Code Administrators and Inspectors Fund, with any excess going to fund the Florida Homeowner's Construction Recovery Fund). Payouts per violation are capped at $50,000 per homeowner, and the cap against each contractor is $500,000.

Only specific events are covered, such as the builder abandoning a job for more than 90 days; causing financial harm to a customer by misconduct or mismanagement; or, signing false statements saying that work is bonded, that all payments to subcontractors have been made, or falsely claiming to have provided proper worker's compensation and insurance.

For more information or to request a claim form, call the Construction Industry Licensing Board at 850-487-1395 or visit: http://www.myflorida.com/dbpr/pro/cilb/cilb_index.shtml

Insulation Disclosure for New Construction Homes

If you are buying a new construction home in Florida, you must be given an Insulation Disclosure, either in the contract or as a separate rider attached to the contract. The type, thickness, and R-value of the insulation in the interior walls, exterior walls, and the ceilings in all areas should be disclosed to you. To learn more about R-Value, see Chapter 13, New Home Construction.

A disclosure may look something like this:

Location	Type	Thickness	R-Value
Ceiling/Batt	Fiberglass	10"	R-30
Ceiling Blown-In	Fiberglass	13"	R-30
Exterior Walls	Foil	-	R-4.1
Interior walls/Batt	Fiberglass	3"	R-11

10.
HOME INSPECTIONS AND WARRANTIES

Home inspections and home warranties are two tools available to you that will help to ensure that the present and future condition of your new home here in Florida is satisfactory.

HOME INSPECTIONS

Before finalizing the purchase of a home, you should always have a home inspection done. This point cannot be stressed enough. A home inspection could be the best money you ever spend. Home inspectors conduct a thorough evaluation of the home that can help you understand the condition that the house is actually in before you take ownership. Remember, that sometimes looks can be deceiving, and nobody likes unexpected surprises or costly repairs once they move in. Even if the house is fairly new and appears to be in good condition, you never know what could be hiding out of plain sight.

Most home inspections don't reveal much of anything, maybe faucets that need tightening, or caulking that needs to be done. The point of an inspection is not to convince you that so much is wrong with the house that you are discouraged to buy it. It is rather to give you an accurate depiction of the current condition of the house, as well as an idea of how certain things will hold up in the future.

When major items are found, such as a failing air conditioning unit, or bad wiring, the parties must look to the real estate contract to see who will be required to make the repairs. Home inspections typically cost anywhere from $200 and up, depending on the size of the home. After the inspection is complete you will be given a detailed report of all the inspector's findings, whether good or bad, usually accompanied by digital photos.

A typical home inspector will inspect the structural elements of the home consisting of the roof, outside and inside walls of the home, patios and driveways, as well as parts of the foundation if visible. They will go into the attic to inspect the trusses, the underside of the roof decking for water intrusion, and insulation. The systems of the home will be inspected including the electrical, HVAC, and plumbing systems. All appliances that are staying with the home are inspected and tested for proper operation, and usually a random spot check of electrical outlets, windows, and doors will be done.

Other items that a home inspection company might perform for additional fees include radon gas and mold testing, water analysis, and pool and spa inspections. Most home inspectors subcontract for a termite inspection that may be at an additional cost to you, but it is a very important part of any home inspection here in Florida.

Some national home inspection companies with branches in Florida for you to look into are:

Amerispec
http://www.amerispec.com

Pillar to Post
http://www.pillartopost.com

Many local home inspection companies perform just as well and may have specialized knowledge of the homes in the area. Ask your real estate agent, or friends and family for referrals.

HOME WARRANTIES

If you are buying a new home from a builder, one of the advantages you have is that your home will usually come with a warranty provided and paid for by the builder. But this doesn't mean that you're out of luck if you decide to buy a resale home. There are several home warranty options available to you, no matter the age or condition of the home you are buying.

Home warranties for average homes under approximately 5000 square feet will cost you between $300 and $400 dollars per year. You can renew these on a yearly basis. Most plans do not require an inspection of the property before they take effect. Depending on the company you choose and the specific plan you go with, an additional amount may be needed to warrant some items like the A/C, refrigerator, washer/dryer, and a pool or spa.

Typically covered items include the plumbing, electrical, and heating systems, water heater, most appliances, disposal, smoke detectors, and exhaust fans. You need to read the warranty contracts carefully to see exactly what is and what is not covered. For example, a warranty might cover your refrigerator motor, but not the shelving inside the refrigerator.

Should something that is covered by the warranty break down, there is usually a service call fee, anywhere between $40 and $80. Other than paying that, you will not be required to pay out any money for the repair or replacement of a covered item. Most home warranties are pretty simple to acquire, fairly inexpensive, and are usually worthwhile.

Some national home warranty companies to consider are:

American Home Shield
http://ahswarranty.com

Old Republic Home Protection
http://www.orhp.com

11.
NEGOTIATING

With most purchases we make in our daily lives, negotiation is usually not part of the transaction. We don't try to haggle with the barista at the corner coffee shop over the price of our grande white mocha, no whip, with non-fat milk. Because of this, when we go to purchase big ticket items like cars and homes, we are often times reluctant to negotiate because of our natural desire to avoid conflict. However, when purchasing a home, negotiating is, most of the time anyway, the order of the day. I say "most of the time" because if you are buying a new home in a community directly from a builder, you may not be able to negotiate. Most builders do not negotiate, or at least don't like to tell people they negotiate.

But that can depend on the market. For example, if in a community you are considering, a builder has several homes completed with no buyers for them, you may be able to negotiate with them for something. It may end up being tens of thousands of dollars, it may be a couple hundred dollars, or it may be a free refrigerator. It just depends. But as the saying goes, "you never know, unless you ask!" You just don't want to end up paying full price when you could have saved thousands or more just by asking.

IT'S NOT JUST ABOUT THE PRICE

Better prices aren't the only thing up for grabs. When we hear the word "negotiation" the first thing we think of is price. But real estate transactions are so complex that there are many things that can be negotiated other than just the price. You can negotiate things such as when the closing will take place, who will pay for any repairs needed after the home inspection, who will pay the transaction closing costs or attorney's fees, and who gets the patio furniture. The most important thing for you to remember is that almost everything is negotiable.

TOP FOUR TIPS FOR NEGOTIATING

While no two real estate negotiations are exactly the same, you will find yourself in a position for success nearly every time if you keep these four simple negotiating tips in mind.

Keep your emotions in check.

The main thing to remember in any negotiation is to remain calm. It can help if you have a real estate agent representing you to act as a buffer between you and the seller or the seller's agent, so long as the lines of communication are clear and open on both ends. Keep in mind that this transaction is probably just as stressful for the seller as it is for you. I recently sold a personal home of mine, and even though I'm a seasoned real estate broker, I let another broker I trust handle my sale because its just too tough to be objective when your own emotions are involved. You do not want to come out of this transaction thinking you paid too much, and the seller does not want to feel like they "gave the house away." This segues into my next tip...

Do your research before submitting an offer.

If you are working with an agent, ask them for a list of homes comparable in size, style, age, and location to the one you are considering (this is called a CMA or Comparative Market Analysis) and compare the sales prices of those homes to the price of the home you are considering. If you are not working with an agent and do not have access to the Multiple Listing Service, the local property appraiser's website usually will have data such as sales prices online. Another new resource emerging for this task is the website Zillow.com which I mentioned earlier. Keep in mind that these websites sometimes take anywhere from 6 weeks to six months longer to update than the Multiple Listing Service, and therefore may not have the most current information.

Set your limits before you submit an offer and be realistic.

If all you can afford to pay or are willing to pay is $300,000, you are not doing anyone any favors by looking at $400,000 homes and offering $100,000 less and hoping someone will bite. By the same token if you want or need a four bedroom home but get talked into or decide to settle on a three bedroom home, you may be unhappy down the road when you realize you really did need that extra room. Have a firm understanding beforehand of what your wants, needs, and limitations are and stick to them.

Be the path of least resistance.

If roles were reversed and you were selling a home and were placed in the following two scenarios, which buyer would you like to have most?

Buyer "A" has been pre-qualified by a lender, has asked for no contingencies other than the right to inspect the property and have repairs made, and wants to close within 30 days.

A lender has also pre-qualified buyer "B," but they want to make the contract contingent upon their current home selling within 45 days. Both buyer "A" and buyer "B" are offering the same amount of money.

Most likely you would choose to accept the contract from buyer "A" and reject or at least counter buyer "B" with terms more advantageous to your position. It looks like Buyer "A" is financially able to purchase your home and they don't want to wait. Buyer "A" wins over buyer "B" every time.

The lesson here is that when you are searching for a property in a fast moving market, be buyer "A" and give the seller every reason to choose you and your offer over anyone else's. Be the seller's path of least resistance.

12.
REAL ESTATE FINANCING

Real estate financing in Florida, for the most part, is no different from real estate financing in any other state. However, if it has been a few years since you last went through the process of getting a mortgage, you could use some refreshing on the world of real estate finance.

TYPES OF LENDERS

While shopping for financing during your home buying process, you may encounter several types of mortgage lenders. Here is an explanation of some of the more common types of lenders you may encounter along the way to homeownership.

MORTGAGE BROKERS

Mortgage brokers are essentially middlemen in the mortgage process. They serve as intermediaries between lenders and borrowers. Mortgage brokers typically have the widest array of loan programs available, because they have relationships with several lenders. The term "middleman" is usually associated with extra expense, but mortgage brokers can actually save you money by helping you comparison shop and find the best mortgage for your particular situation. Mortgage brokers will assist you with things such as filling out and submitting your loan application, running your credit report, scheduling an appraisal, and helping to coordinate your closing.

However, the mortgage broker does not make the decision to fund your loan. A person called an underwriter, who is employed by the lender, makes that decision. Mortgage brokers are paid a fee for their services, sometimes charged to the borrower in the form of points or origination fees, but they can also be paid by the lender, or often times a combination of the two. Sometimes this fee can be negotiated in your favor, or you can also ask that the broker just charge a flat fee, a strategy that is recommended by many industry experts.

Mortgage Bankers

Mortgage bankers are in the business of originating loans and then selling them to the secondary mortgage market (more on that in a moment). Mortgage bankers often have appealing loan rates and programs, but oftentimes do not have access to as many sources as the previously mentioned mortgage broker.

Banks and Credit Unions

Most banks and credit unions also offer mortgages. Funds for these loans are obtained from their customers through checking and saving accounts as well as certificates of deposit. The bank or credit union will sometimes service the loan themselves (if it's a large bank) or sell the loan to the secondary market.

The Builder's Lender

If you are buying or building a new home, your builder may also own a mortgage company or could possibly be affiliated with one. They will usually try to get your business by dangling low initial interest rates or credits toward closing costs in front of you. Whether or not going with the builder's lender will pay off for you in the long run is not a simple question to answer. But here's my advice for what you can do to try and figure it out.

First, get a good faith estimate from the builder's lender. I'll discuss good faith estimates in depth a little later, but for now, just know that it is an estimate of how much money you will have to pay out of pocket at closing, what your closing costs will be, and what your monthly payments will be. With this in hand, take it to a mortgage broker not affiliated with the builder or builder's lender and have them explain to you the good and bad points of the builder's financing options. There is no sense in trying to dissect it on your own. There are so many places that lenders can hide fees it'd make your head spin.

A good mortgage broker won't mind taking the time to explain the pros and cons to you, because they ultimately hope to gain your trust and subsequently your business. However, if the deal you are quoted by the builder's lender is a good one, an ethical mortgage broker will tell you so. If he can't meet or beat the builder's financing, you might want to go with the builder's lender.

SECONDARY MORTGAGE MARKET

It is common for lenders that provide home loans to sell these loans to the secondary market, made-up of investors such as Fannie Mae and Freddie Mac. Selling your loan provides lenders with the funds they need to issue new mortgages. If your loan is sold, it will not affect the terms of your mortgage or your payment. It will however affect who your payments are made payable to, so if you are using an online bill payment method make sure you are paying the correct entity and sending your payments to the correct address.

TYPES OF MORTGAGES

Just as there are many sources for your new mortgage, there are also several different types of mortgages. Here are some of the most common types of mortgages.

Fixed Rate Mortgage

The most common type of mortgage, a fixed rate mortgage, is one in which your interest and principal payments remain the same (constant) over the life of the loan. Bear in mind that your total payment may fluctuate (usually upwards) as real estate taxes and homeowner's insurance rates change over the life of your loan. Different terms are available for fixed rate loans, from as short as 10 years to new 40 and even 50 year mortgages, loan periods which were recently introduced. Keep in mind that the shorter the term, the higher your monthly payment will be. However, the longer the term of the mortgage, the more interest you will end up paying over the life of the loan.

Adjustable Rate Mortgage (ARM)

Adjustable-rate mortgages are mortgages in which the interest rate on the mortgage fluctuates over the life of the loan. The rate will initially be fixed for a specified period of time. For example, with a 5/1 ARM the rate will be fixed for 5 years and adjust every year after that. Rate adjustments are made based on changes to a defined index. The interest rate is determined by adding a fixed number of points to the index. The attraction with adjustable-rate mortgages is that rates are initially lower than that of fixed-rate mortgages. If you do not intend to live in a house for longer than the initial fixed rate period, you will not be subjected to the adjustments in the rate. The disadvantage of an ARM is that during times of rising interest rates, your payments can increase dramatically after the fixed period is over.

Balloon Mortgage

A balloon mortgage is a short term mortgage, usually 2 to 7 years in length, that is amortized over 30 years with the balance becoming due in a lump sum at the end of the term. Again, rates are lower than fixed-rate mortgages, but some people do not want to have to refinance or pay a large lump sum at the end of the loan term.

Reverse Mortgage

Reverse mortgages have been around for many years but are just recently gaining notoriety. A reverse mortgage is a mortgage where the lender pays you either one lump sum or a smaller amount each month, as opposed to you paying them. This can give you extra money to pay your bills and do the things you want to do but otherwise might not be able to afford.

When you pass away or decide to move, your heirs or new owners get ownership of the home and must repay the mortgage. This is most commonly accomplished by selling the home. Reverse mortgages are only available to people over the age of 62, and should only be considered in specific circumstances. In fact, before obtaining a reverse mortgage you must be counseled by an HUD approved reverse mortgage counselor.

VA Loan

Veterans of the United States Armed Services with more than 180 days active duty during peacetime, or 90 days during times of war may be eligible for a VA loan through Uncle Sam. VA loans can be used to purchase a home, manufactured home, or condo. In order to obtain a VA loan, the law requires that:

- the applicant be an eligible veteran who has available entitlement
- the veteran must occupy or intend to occupy the property as a home within a reasonable period of time after closing
- the veteran must have satisfactory credit
- and, the veteran and spouse must show stable income sufficient to meet the mortgage payments.

The advantages of VA loans are that they require no down payment, they are available from most lenders, and the VA prohibits lenders from requiring PMI, or Private Mortgage Insurance. The VA is guaranteeing the loan, so there is no need for a lender to require the veteran to pay for additional insurance against default.

On the downside, VA loans carry a one time funding fee ranging from one and a quarter percent to three percent, depending on the veteran's service, as well as other factors.

> For more information on VA loans, visit www.va.gov or call the Florida VA Regional Loan Center at 1-888-611-5916.

SPECIAL FINANCING

There may be times when a conventional real estate loan will not meet your needs. For these cases, special types of real estate financing may be available to you.

B/C Credit Mortgages

B/C loans, sometimes referred to as sub-prime loans, are for those who do not meet the credit guidelines established by Fannie Mae or

Freddie Mac. Through B/C loans, borrowers are able to obtain financing for a temporary period of time, when their credit history excludes them from receiving "normal" loans. Recent bankruptcy, divorce, foreclosure, or late payments that show up on your credit report can bump you into this category. You should be aware though that loans of this type carry higher interest rates than those of "A" credit borrowers.

Bridge Loans

Just as Citizen's Property Insurance is the insurer of last resort, a bridge loan should be your "financing of last resort". Let's say you look at a home in Florida, fall in love and have to have it. But, you haven't yet sold your home in New Jersey (or wherever you're from). A bridge loan will allow you to purchase the new home in Florida without having to first sell your home up North. There are essentially two ways for a bridge loan to be structured.

The first way is you get a bridge loan for enough money to pay off your current home and make your deposit on the new home in Florida. Then you would just get a regular mortgage on the new home. You won't have to make payments on the bridge loan for a predetermined amount of time, say 6 months or a year, but in the meantime, interest is accruing. The rub lies in that if you don't sell your home in the allotted time, you will have to start making payments on the bridge loan, meaning you'll now be making two mortgage payments. Once your home up north sells, you pay off the balance of the bridge loan and any interest that has accrued.

The second way to structure a bridge loan is to use the equity in your home up north to make the down payment on the home in Florida. Now you have two loans, your original mortgage, and a second home equity mortgage. Then, you'll get a mortgage on the new house. So, essentially you have three loans. But, you aren't usually expected to pay on all three, just your original mortgage and your new mortgage. Again, once your home sells, you'll pay off your original mortgage and the bridge (second equity mortgage) as well as any interest that has accrued.

Experts only recommend getting a bridge loan if you know that you can afford to make two mortgage payments if you had to. Usually, a lender won't give you a bridge loan unless you have enough cash to make both payments anyway. Also note that the rates on a bridge loan will be significantly higher than say, a typical 30-year mortgage. It can pay to shop around to different lenders to see what types of bridge loans they may offer and what the rates are.

INTEREST RATES

News of interest rates is everywhere, your local newspaper, online and on television. Some people in the real estate and finance worlds hang on every eighth-of-a-point fluctuation in interest rates. For most of us, however, there is little need to do this. A quarter-point here or half-point there shouldn't affect your plans for purchasing a home in Florida, so long as you have planned wisely in the financial department. For average size mortgages these fluctuations won't add but a paltry sum to your mortgage payment. But it is important to know a few things about interest rates, in hopes of better understanding how they could affect you.

How Interest Rates are Determined

Mortgage interest rates, contrary to what many people believe, do not follow the Federal Reserve Board's lowering and raising of rates. Instead, they actually anticipate the fed. A closer tracking device for mortgage interest rates is the 10-year Treasury note. If you want to know what mortgage interest rates are doing, follow the 10-year Treasury. But as mentioned before, slight changes in rates are nothing to lose any sleep over.

Rate Locks

The typical escrow (time from contract to closing) on a home is 30-60 days, but interest rates are constantly changing. In order to protect yourself in an environment of rising interest rates, get a rate lock. With a rate lock the lender holds or guarantees the interest rate for you for a predetermined length of time. Sometimes they will do this for free for a minimal amount of days, say 15-20, and for longer periods they will charge you a fee. A 60-day rate lock will be more costly than a 30-day rate lock.

Sometimes during escrow, rates will drop, leaving you paying a higher rate than the market rates at time of closing. To avoid this, ask for a rate lock with a one-time float. If the rate goes lower anytime before closing, you can float down to the lower rate. Again, some lenders offer this feature for free, with others you will have to pay. Shop around because the market is always changing and so are lenders' terms and policies. Competition among lenders can run high, especially in slow markets.

Interest Rate Buy Downs

An interest rate buy down is a reduction in the interest rate that you pay on a mortgage. There are temporary buy downs and permanent buy downs. Temporary buy downs are common as an incentive for builders; they pay the lender a fee to get the buyer a lower initial rate for a set period of time, usually a year or two. This is also an incentive sometimes offered by home sellers to entice someone to choose their home over another.

As a buyer, you can also acquire a permanent buy down. With a permanent buy down, you pay a fee up front to have your interest rate lowered or "bought down" for the life of the loan. You should only do this if you plan on keeping the mortgage for a long time, as it will take a while for the lower rate to recoup the money you paid out to buy it down. Check with your mortgage lender to see if a buy down might make sense for you.

APPLYING FOR A MORTGAGE

The mortgage application is going to be mostly what you would expect. It will ask for your name, social security number, your address for the past two years, a copy of your driver's licenses, a list of all your assets as well as debts along with monthly payments, employment information, sources and amount of all your income, and more. The type of lender and type of loan you choose will determine the additional information that the lender will ask you to submit along with your mortgage application. These can either be faxed, e-mailed, mailed, or hand delivered to the lender, depending upon your location and theirs. Just to give you an idea, some of the items they may need include:

- a copy of your sales contract (if you've executed one)
- proof of your deposit (copy of the cancelled check, bank statement, etc.)
- pay stubs for the last 30 days if you are still working
- your past two years of tax returns
- statements for all your bank accounts and investment accounts for the past three months

REAL ESTATE FINANCING

- a copy of your current mortgage statement if you have one
- if you are divorced, they will probably ask for your divorce decree
- if you are self-employed they will need a current profit and loss statement
- if you receive a pension or social security the lender will ask for proof of such

Again, depending on the lender and type of loan you are applying for this list can be longer or shorter.

GOOD FAITH ESTIMATE

Within three days of applying for a loan, you should receive a "good faith estimate" as well as a HUD guide to settlement costs from the bank or mortgage company that you applied to. The good faith estimate is just that, an estimate of the costs that you will be expected to pay at closing. Costs that will be reflected on the good faith estimate will include costs for appraisals, surveys, attorney's fees, recording and transfer fees, doc stamps, mortgage origination fees, and more. Never fully commit to a lender until you have reviewed and are comfortable with their good faith estimate.

1031 EXCHANGE

Instead of getting a mortgage to pay for your new property in Florida, why not trade for it? You may already be familiar with Internal Revenue Code Section 1031, that allows you to defer any capital gains on the sale of an investment or income property by investing those gains into another "like-kind" property. Most commonly referred to as "1031 Exchanges" or "Like-Kind Exchanges" these complex transactions, when executed properly, can help you purchase your eventual retirement, investment, or second home property in Florida. You should of course consult your tax advisor before considering this type of transaction to see if it supports your financial, investment, and real estate purchasing goals.

Should you get the green light to proceed, your next step should be to contact a Qualified Intermediary. Ask your tax advisor, accountant, or real estate agent for referrals. A Qualified Intermediary is a company that specializes in 1031 Exchanges. Once you contact them, they will guide you as to how to structure any sale or purchase contracts to facilitate the exchange. You will have specific time constraints to adhere to, namely 45 days to identify a property you wish to exchange for, and 180 days to close.

In a perfect world, you would close on the property you are giving up or selling first, then purchase a new property. This is called a delayed exchange, and is the most common, least expensive, and least paperwork intensive type of 1031 exchange. However, as we all know, the world is not always perfect, and sometimes the opportunity to purchase a prime property will pass you by if you do not act in a timely fashion. You may find it necessary to purchase a property before selling your current property. Most people do not know that "reverse exchanges" are available that will allow you to do this and still defer your capital gains. It will be at a higher cost and with much more paperwork involved, but it can be done.

> For more information on 1031 Exchanges, contact your tax advisor or accountant. To find a qualified intermediary visit http://www.starker.com.

YOUR CREDIT SCORE

Most everyone has heard of a FICO score, created by Fair Isaac Corporation. In case you haven't, it's one score that lenders will look at to determine how good of a credit risk you are, and consequently how much money they will lend you and with what terms. FICO scores range from 300 to 850 and the median score nationwide is 723. If your score is above that median, you are doing pretty well as far as most lenders are concerned and you should qualify for the best rates when shopping for a mortgage.

The big three credit reporting agencies, Equifax, Experian and Transunion also produce individual credit scores based on the information they have about you and your history. The problem, however, with credit scores is that they are generally unpredictable, and can vary widely from different reporting agencies.

The problem that a lot of people face is that they have no clue going in to apply for a mortgage what their credit score is and when they find out it's too late to do anything to improve it. Experts recommend that at least six months before applying for a loan you should visit www.MyFico.com to get your score. There is a fee involved with this but the knowledge you will be armed with after finding out your score could prove to be invaluable.

Ways to Improve Your Score

There are several ways to beef up your credit score in the months leading up to applying for a loan. Most important, experts agree, is to keep your credit card balances below 25% of what your available credit limit is. Thought you might escape those library fines since you're leaving town? Not so fast. If they are turned over for collection, they can damage your score, so make sure you are all square with the house. Also, do not open or close any credit accounts, including car loans, in the time leading up to applying for a loan. Both can hurt your score.

YOUR CREDIT REPORT

Your credit score is based upon the information that can be found in your credit report. Everyone is entitled to a free copy of his or her credit report once a year. Simply visit www.annualcreditreport.com to get a copy. Be sure to check your report for any errors, such as erroneous bad debt claims and the like, which can drag down your score.

> For more information on credit scores and credit reports visit:
> http://www.myfico.com
> http://www.annualcreditreport.com

MORTGAGE CALCULATORS

> Visit http://www.floridaforboomers.com and click on "Mortgage Information" and then click on "Mortgage Calculators" to see what different payment amounts are for the type of mortgage you are interested in.

13.
NEW HOME CONSTRUCTION

For some boomers, building a new home in Florida is the ultimate dream. There's not much more exciting in real estate than picking a lot, choosing or designing a floorplan, making your interior design selections, and seeing it all come together right before your very eyes.

Besides being exciting, it can also be described as overwhelming, daunting, mind boggling, and a slew of other adjectives. After reading this chapter, you should feel comfortable enough with Florida new home construction that if you do choose to build a new home here, the words you use to describe it will hopefully lean more towards "exciting."

CHOOSING A BUILDER

There are many factors that should go into your choice of home builder. The most important factors are those that are important to you. Of course, these are different for everyone. Some people want the very best price available, while others don't mind paying more for higher quality. Some want a builder that will hold their hand throughout the entire process, while others prefer to have very little contact with the builder. Others still want total control over their selections, while some don't mind if the builder even chooses the colors. It's very important that you decide for yourself what factors are important to you before you start looking for a builder.

Once you decide on the factors that are most important to you, do some investigating. If it's the best price you desire, visit several communities and see which builder is offering the best incentives. Builders with numerous homes in their inventory ready to move into are more likely to give incentives than a builder who has no standing inventory. Check the local paper in the area you are looking for homes, and scan the advertisements for good deals and incentives. A local real estate agent might be able to point you in the direction of the best deals in town.

If quality is the most important thing to you, spend some time in the builder's model homes looking around on your own and examining things like the trim work, the drywall, and the paint. Look for anything that's not quite right such as wavy walls or uneven paint applications. Chances are if the builder didn't take the time to get his models right, he won't take the time to get your home right. If the builder does not have a model, see if he has a list of customers that you may contact to try and go see their homes.

WORKING WITH THE BUILDER AND HIS STAFF

We all know that birds of a feather flock together. The same typically holds true for builders and the people they employ. The first person you will meet when looking at model homes will be the builder's sales staff. Start with them. Are they presentable, eager to help answer your questions, and polite? Or are they abrasive, pushy, rude, and uncaring? Typically, if they like their jobs, and take pride in the product they are representing, odds are good that you're looking at a pretty good builder.

Remember that the salesperson will typically be the direct line to the builder for you throughout the homebuilding process. Keep in mind that you are probably not their only customer. So if they are busy with others when you stop by to ask a question or report a complaint, be respectful of them and their time. Respect is reciprocal.

Some builders will have you meet with different members of their staff during different phases of the construction process. You may meet with a decorator, an architect, a superintendent, or all of these before and during the construction process. To save time for everyone, before meeting with each representative, have your questions ready.

A few builders allow their customers to walk the construction site whenever they want, others only allow it at specific times during the process. In either case the builder's insurance policy usually does not cover you if you are injured, if you step on a nail, or trip and fall over some plywood. Construction sites, no matter how well they are supervised can be dangerous; so most builders require that you be escorted by someone on their staff when you visit the construction site and that you visit at times when subcontractors are not working inside. It may not always be convenient or possible for you to get to the construction site during the builder's business hours. If that is the case, you are at your own risk when you visit the site.

HOW TO CHOOSE A FLOOR PLAN

Obviously you need to choose a floor plan that fits well with the way you live your life. If you plan to use your home in Florida just as a second home or vacation home, and rarely expect to have many guests accompany you, then maybe a one or two bedroom condo, townhouse, or small home will fit your needs. If you expect more people, or plan to use the home as a primary residence and are accustomed to a much bigger home, obviously you will want to go bigger with a large condo (maybe even combine two adjacent units), a large townhouse, or a single family home.

The key to finding a floor plan that fits your needs is to spend some time in the builder's model homes, if available, and try to envision things such as furniture layouts, traffic patterns, blending of public spaces, such as the flow from the kitchen to the living room, as well as private spaces, such as bedrooms and bathrooms. Place most of your interest in areas that meet your lifestyle needs. For example, if you love entertaining, look for an extra large great room and maybe an open kitchen. If you plan on having lots of visitors—and remember everyone wants to come to Florida—then focus on bedroom sizes. You get the idea.

You may be accustomed to the split bedroom floor plan. These are popular in Florida as well. However, in some developments where the lots are narrower (such as 40, 50, or 60, feet wide) you may find that split bedrooms are rare. Not many builders have found a way to make an efficient split bedroom floor plan for those size lots. Most people, however, find that once they are in a home with the bedrooms on the same side, that they don't really miss the split bedrooms all that much. Split bedrooms are great for families, especially with teenagers who like to crank up their stereos or play their drums. Having the secondary bedrooms on opposite sides of the house from the master bedroom provides parents more peace and quiet. But most baby boomers retiring to Florida or using their home as a second home don't have that problem, and find that a non-split plan works just fine for them.

CHOOSING A LOT

The selection of a lot to build your home on is, for some people, even more important than choosing the floor plan or who builds your house. We all know the saying, "location, location, location."

Some people are very particular about their lot, as well they should be. The lot you choose will determine several things, such as the quality of view you will have, your level of privacy, your utility bills, the price, and your home's future resale value. Some people could care less what lot their home is built on, but taking everything into consideration, they certainly should.

In Florida, water views, be it the ocean, river, lake, or pond are the most coveted, followed by views of a golf course. As such, prices and premiums you will pay to look at these vistas are higher than say, a lot with a view of the interstate. Also many people like to have their home back up to a conservation area, which assures them of privacy in that nothing can be built behind them.

Most people don't think about it but the lot you choose can also have an effect on your heating and cooling costs. On most homes, the majority of the windows are located on the front and rear of the home. If the home is placed on a lot with an east/west exposure (home faces east or west) more sunlight will enter the home, increasing your utility bills in the summer time, and lowering your heating costs in the winter.

There are other considerations if you are going to have a swimming pool at your home. If the rear of your home faces east, your pool and patio areas will be shaded from the sun in the afternoons, but will have the morning sun. Some people prefer the afternoon shade while some prefer the afternoon sun. Others still prefer a north/south exposure, which evens out the two extremes. There is no right or wrong answer, but you'd be doing yourself a disservice if you did not at least consider what you might like before choosing a lot. My advice is try to visit the lot you are considering at different times of the day and see what exposure you think will work best for you.

WHAT IS A ZERO LOT LINE?

You may notice that in some communities here in Florida, the houses are extremely close together, sometimes a few feet or less apart. What you are probably looking at is a zero lot line community. Zero lot line means that the house is placed either on or very close to one side of the lot, allowing for a little more yard on the opposite side, and sometimes also very close to either the front or the back, which allows you to have either a bigger front yard or bigger back yard.

This method allows for more homes to be built in a smaller amount of area, and is great for someone who wants very little yard to maintain, but does not want to sacrifice home size to accomplish that goal. Another benefit is that many zero lot line communities are maintenance free, so as the owner you won't have to worry about cutting the grass anyway.

BE AWARE OF THESE BUILDER CONTRACT CLAUSES

As opposed to using the standard "FAR/BAR" contracts like in the resale side of real estate, builders use their own contracts, and they usually err on the side of protecting the builder rather than the buyer. By making you aware of the contract clauses you might see in builder contracts, the goal is not to steer you away from them, as most will be unavoidable. The goal is to heighten your awareness of them, and the impact that they might have on your new home construction experience. As is the case with any other contract, if there is anything you are unclear about or do not understand, you should consult an attorney before signing.

Here is a selection of some clauses you should be aware of that you may see specifically in new home sales contracts.

Deposit

In most cases, the deposit that you give a builder at the time of contract will be substantially higher than if you were entering into a resale contract. Most builders here in Florida require 5-10% of the sales price as a deposit, and some even require 20% down at time of contract. Now, what happens to that deposit? In a resale transaction, the deposit would be held in an escrow account at a bank, either by an attorney, title company, or real estate broker. In the case of new construction, that money is sometimes used to build your house. If this is the case, the builder's contract will have a deposit clause, whereby you as the buyer waive your right to have the money placed in an escrow account.

Financing

Some builder contracts do not include a financing contingency like the Standard FAR/BAR contract does. This would mean that if you gave the builder a deposit on a home, but you are denied for financing, you could lose your deposit. The chance that this could happen increases the importance of getting pre-qualified by a lender before shopping for a new home to get a feel for what you can afford. In any case, when you sign a contract with a builder, make sure you know whether or not it includes a financing contingency.

Construction Completion

Several factors will determine how quickly your new home is built, such as the market conditions at the time you do a contract, availability of labor, and the builder's backlog. For example in a heated market such as Florida experienced in 2004-2005, homes were sometimes taking 12-15 months to be completed. The main reason for this was there were a large number of homes under construction. If you had bought a year earlier in 2003, your home may have only taken 6 months to complete. Some builders use a construction completion clause to spell out how long the builder has to build the home for you. While it theoretically may only take six months to build your home, some builder contracts allow the builder to take years, yes with an "s", to build your home.

However, don't take this clause to mean that the builder will take the full amount of time to build your home, simply because the contract says they can. Builders want to get homes finished and closed as soon as possible because that's when they get paid. They want you in your new home just as quickly as you do.

Construction Delays

This is a separate clause found in most builder contracts that ties into the previous clause. It basically says that the construction completion can be delayed for reasons that are beyond the builder's control such as acts of God, adverse weather, theft, or a shortage of labor or materials.

Closing Date

Some builders place a firm closing date in their contracts, while others leave it open ended. Builder policies vary, and you will have to ask about the policy of your builder regarding closing dates. Without a firm closing date you can be left in limbo not knowing when to schedule movers, utility hookups, change of address, delivery of new furniture, and so forth. This can understandably be a nightmare. Even if builders do not specify a closing date at the time of contract, most will at least give you a week or two notice when the house is complete to arrange for your closing and subsequent move.

Escalation Clause

Think not having the slightest idea of when your closing will be is bad? What if you didn't know how much your new home would ultimately cost until just before the closing? In order to protect themselves from unexpected rising construction costs while your home is being built, some builders employ escalation clauses in their contracts. These can be worded and structured in various ways but basically they allow the builder to pass on any increase in costs during construction to you, the lucky homebuyer.

If you sign a contract that has an escalation clause in it, at least make sure that there is a cap on how much the price can go up, and that you are comfortable with that amount. This cap can either be a hard dollar amount or a certain percentage. Without this cap, you are essentially handing the builder a blank check.

Termination of Sale Clause

This clause allows the builder to be released from their contract with you at any time if certain conditions occur. These conditions usually include the builder being unable to get the proper permits for your home, being unable to deliver clear title, or for any other reasons beyond their control.

Holding Period; Builders Right to Equity

Keep in mind that 2004 and 2005 were the hottest years on record in Florida real estate. Some real estate investors referred to as "flippers" were buying homes from builders and selling them just a few months later for big profits, as home prices in the state skyrocketed. Different builders had different names for this contract clause and only a very small percentage of builders put it into their contracts.

In a nutshell, this clause told a buyer 1) how long they had to own the home the builder was selling to them before they could turn around and sell it and 2) that if they sold before that time period was up, the buyer had to share a percentage of the gain on the sale with the builder.

The idea behind the clause was that if a builder had, say, 400 homes in a subdivision to sell, the clause would limit the amount of competition they would have from investors who bought early on in the project and might decide to sell quickly and take their profits by undercutting the prices of the builder.

It's not likely you'll see this clause in builder contracts in the very near future, as price gains have slowed dramatically since then. But that's not to say that if you buy in the coming years that you won't see it. Prices are expected to soar again in the next 5-10 years and builders may pull this clause out of their filing cabinets and dust it off to use again.

MAKING YOUR SELECTIONS

After finalizing your contract with the builder, you will be given a time to either meet with the builder's decorator, or to visit their design center. This is when you pick all the colors and interior materials that will give your home its personality, including things like cabinets, carpeting, tile, countertops, paint colors, and so on. Depending on the size home you've chosen, this may be a short two to three hour process, or it could span several appointments over the course of a couple days or weeks.

No matter how many homes you build, this will always be one of the most stressful times because what you choose here will determine how your home is going to look and function for years to come. To minimize the time and stress that picking everything out will invariably cause, it pays to have done your homework as to what you like beforehand. This way you won't be making any split second decisions on things that you might not be able to change later. Take pictures of ideas you see in model homes or tear out pictures in magazines of things you might like to have in your new home.

CRITICAL STEPS IN THE NEW HOME CONSTRUCTION PROCESS IN FLORIDA

Have you ever bought a perfect car? Wait a second—I know you're thinking—did I pick up the wrong book? I thought this was the book about Florida real estate written by that tall, handsome gentleman. Yes, you've got the right book.

For most people, the answer to this seemingly out of place question is no. No matter what make, model, or how much you pay, there seems to always be at least one problem or imperfection that you notice within the first couple weeks of owning a new vehicle.

Well, no matter where you purchase or how much you pay, don't expect anything different with a new home. Just think, cars are built in a factory, protected from the elements like sun, heat, and rain. Machines and computers also play a big role in the production of a car.

Unfortunately, homes don't have those advantages. They are built in the dirt out in the hot sun, being rained on and rained in, often exposed to the worst Mother Nature has to offer. And to be quite honest, a high school degree is a major accomplishment for some of the people doing the actual physical labor on your home. Sure, some may even have some college under their tool belts but remember, digging footers, laying block, and hammering nails aren't $50 an hour jobs.

Fortunately for you, the builder supervising them in most cases is well educated and more importantly licensed to make sure your home is built up to the requirements of the city, county, or municipality where the permit was issued. You've also got city inspectors who are trained to discriminate between good work and bad work, looking out for you.

And finally, you have the next several pages of critical steps in the new home construction process to help ease your mind during what can be a confusing and frustrating time. While it won't by any means tell you how to go out and build your own home, this information should give you the knowledge to feel a bit more comfortable with everything that will go on during the new home construction process.

Permitting

Once the floorplan and other structural features of the home have been selected, the builder will submit the plans to the city for approval. When submitted, the plans for the home must usually be accompanied by an architect's or an engineer's seal, essentially stating that they certify that the home is planned in accordance with the proper design specifications and building codes. Hard construction cannot begin until the permits have been received back from the city.

If the city feels the plans submitted comply with all current zoning and building codes, they will issue the builder permits to build the home. A copy of the plans and permits will be kept in a permit box in plain view at the construction site and checked frequently by city inspectors throughout the construction process.

Fill, Compact, and Site Prep

Most of the residential lots being built on these days in Florida require at least some amount of fill dirt to bring them up to the elevation required by the city or municipality which issued the building permit. Usually once the lot is filled and compacted, an engineer will come out and conduct a compaction test, to make sure the dirt added to the lot has been properly compacted. Not every lot that is filled needs to be tested for compaction. The builder usually has a set standard for lots that they test, such as lots requiring more than one foot of fill. A lot that has not been compacted properly is prone to settling, which can cause cracks in foundations and walls, and more trouble for you down the road.

Anything else that stands in the way of construction of your new home will be removed at this point. Brush will be cut back. Trees too close to the home are susceptible to damage by trucks and other machinery, and tree roots can damage the foundation, so if any trees still need to be removed it will happen at this time. The lot will be graded, generally sloping slightly towards the front, back, and sides to help with drainage. The area of the lot where the home will go will also be leveled, so that in the end, your house will be level as well.

Hub and Tack

Once the lot is filled and compacted, a survey crew will come and stake out the home, also called hub and tack. At this point, the corners of your home will be set.

Form the Slab

Pieces of lumber, typically 2 x 10s turned on their side, are used to create the perimeter of the slab. Footers are then dug out underneath the 2 x 10s. Footers, which are a couple of feet deep (depths vary area to area and builder to builder), and wider than the walls of the home, provide the support necessary to help make the house more sturdy.

Metal reinforcement rods will run around the perimeter of the home and will be positioned vertically at certain intervals to go up inside of the block that will be placed on top of the slab. These will eventually help connect the slab to the tie beam at the top level of the block.

Elevation and Setback Survey

At this point, a survey is done to make sure that the home is being built within the confines of its particular lot, and does not encroach on neighboring properties. The elevation of the lot is also checked, to be sure that it is at the height required by the city, county or municipality that issued the permit.

Rough Plumbing and Inspection

Water and sewer lines, which will be in place under the foundation of your home, are run at this time. The water lines will typically be made of copper, and drainage lines will be made of PVC piping. Any

NEW HOME CONSTRUCTION

electrical outlets needing to be placed in the floor can also be installed at this time. Otherwise the slab will have to be cut later to install them. Once complete, an inspector will verify that each element of the rough plumbing has been installed properly.

Termite Treatment

Most builders in Florida will provide some form of termite treatment for the home. The most common type is a slab pretreatment that will be done before the slab is poured. The purpose of this pretreatment is to prevent termites from getting into your home and doing serious damage in the future. If you were to visit your new home the day that it is treated for termites, you may be overcome by a very strong odor. That's the termite treatment.

Due to environmental, scheduling, and cost concerns, some builders will not pretreat your slab but will treat the actual wood inside your home. After the framing is complete you may see that it looks like the bottom three or four feet have been stained, usually a greenish color. In this case, a termite treatment such as Boracare® has been used to treat your home for termites. The jury is still out as to which is the better treatment for the prevention of termites.

Prep the Slab

Once the rough plumbing is completed, a vapor barrier comprised of several sheets of plastic will be placed over the area in which the slab will be poured. This helps to keep moisture in the ground from penetrating the foundation and getting into your home once the slab has been poured over it.

Slab Inspection

Before the slab is poured an inspector will come and make sure that all work done up until this point is up to code, that all procedures have been properly followed, and construction is safe to proceed.

A side note here about inspections. While absolutely necessary, they can add a significant amount of time to the construction of your home, especially in areas where there is a lot of construction going on. Inspection departments are notoriously understaffed and overburdened with work. Sometimes you may see your house just sitting idle, with no work going on and your natural reaction may be to get angry with the builder.

Believe me, the builder wants to complete your home quickly, sometimes more quickly than you may even want him to. While your home is under construction he is likely paying carrying costs such as a mortgage on the land, insurance, and taxes, not to mention hard construction costs. Understand that when your home is sitting idle that sometimes it is the builder experiencing delays, but most often he is probably waiting on an inspection to be completed before he can proceed.

Pour the Slab

Next, the slab and the footings are poured. In the case of a monolithic slab, one long continuous pour of concrete is all it takes to create your slab. Wire mesh, or more commonly these days, high strength fibers, are usually embedded in the slab to increase its strength and help minimize cracking.

Ideal weather conditions for the pouring of your slab are that the weather should be dry, with little to no chance of rain during the pour, and temperatures should not be extremely hot or extremely cold. If after the slab is poured, it appears rain may be in the forecast, sheets of plastic should be placed over the freshly poured slab to keep it from getting wet. Excess moisture can affect the appearance of the concrete, as well as the integrity of the slab.

While the slab won't reach it's ultimate strength for 20 to 30 days, it will usually be strong enough for construction to proceed in just a few days.

Slab Cracks Eventually with almost every concrete slab, you may see some hairline cracks. They most likely won't appear for a couple months but inevitably some will appear. These do not indicate that you have a bad slab, but are most likely just settling or expansion cracks, the result of extreme temperatures, wind, and evaporation of water in the concrete. One concrete company representative said that the only guarantee they can give is that all concrete will crack, it's just a matter of when and to what degree.

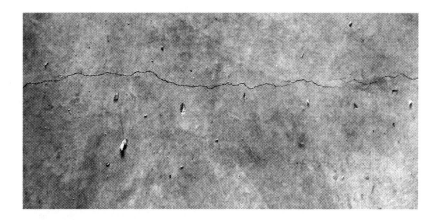

If you are overly concerned about cracks that have appeared in your concrete slab, you might request that the builder hire an engineer to come take a look and certify that it is okay. You will likely have to pay for this service, but if it helps you sleep better at night, it may well be worth the few hundred bucks.

Foundation Survey

Once the slab is in place, another survey will be conducted, just to be extra certain that your home will not encroach on anyone else's property. It's better to discover this problem at this stage, when it will likely cost only several thousand dollars to fix, rather than later when it's a much more costly problem to fix.

Block

It takes a few days for the slab to cure long enough to support block being placed on top of it. Once the slab is cured, the block is then placed on top of the slab to form the outside walls, or shell of your home. Holes are left where the doors and windows will go, and any glass block that you have ordered is normally installed at this stage as well.

You will also see holes every so often along the bottom row of block called inspection ports where the metal reinforcement rods are sticking up from the slab into the block. These holes are included so the inspector can see that the slab, by the use of the metal in it, is effectively attached at set intervals by metal and concrete to the block walls.

Solid Pour Cells, Lintels and Tie Beam

Steel and poured concrete will be added to the block at predetermined intervals, usually every four or five feet. This process is done to add extra strength to the shell of your home and as previously mentioned, connect the shell to the slab. Lintels around all windows and doors will be poured to add strength around these openings, and then the tie beam (top layer of block poured solid) will be poured. This will have steel rods embedded in it as well, and hurricane straps that will later be attached to the trusses will also be embedded in the tie beam.

Lumber and Trusses Get Delivered

About this time in the construction process, your roof trusses and the lumber for your roofing and interior framing are usually delivered. In some cases these materials may sit unused for some time while other projects are being worked on. Just how long they sit out, exposed to the elements, rain, sunshine, and heat depends on the builder's efficiency in scheduling as well as availability of work crews. To minimize weather exposure most builders will at least cover the materials lying out with tarpaulins or plastic sheets.

Trusses and Roof Decking

Your trusses, which will arrive at the home site already assembled and ready to install, act as a sort of skeletal system for your roof.

They will be set on top of the block usually with the assistance of a crane. Once they are secured to the block walls with the metal "hurricane" straps embedded into the tie beam, the roof decking will be placed on top.

OSB vs. Plywood Some builders use plywood roof decking while others use oriented strand board, or OSB. There is endless controversy about which is the better product. The fact is though that both products are excellent in their purest, undamaged forms. The problem that sometimes occurs with OSB is that its exterior "seal" may get damaged on site and lessen its integrity and ability to repel water. Don't fret if your builder insists on using OSB; just make sure that each piece being placed on your roof is in good condition.

Framing

Framing of the interior walls of your home will also be done at this time, and rooms will really start to take shape. Don't worry too much if the framing work looks a little rough at this point. There will be a framing "punch-out" later where everything that needs correcting will be taken care of. Feel free to point out anything that you think may not be obvious to the builder, mistakes can and do happen, but also feel secure in knowing that there will be a framing inspection to make sure everything has been done safely and

correctly. Some builders will also do a framing walkthrough with you just to make sure everyone agrees nothing has been overlooked.

Window Installation

Windows will be installed and usually any sliding glass doors in your home will be installed as well. Windows on most production homes in Florida will be single glaze windows, also known as single-pane windows. If you are coming from a cooler climate you may be shocked at first that builders in Florida use single-pane windows, as you wouldn't have even thought of using them up north. However, single-pane windows are used quite frequently in Florida. One reason is that Florida does not experience the wide temperature swings like the north does. Another argument for them is that it takes about 5-7 years of energy savings to recoup the cost of installing the double-pane windows versus the single-pane, while most people move every 4-5 years anyway.

This doesn't mean you should necessarily settle for single-pane windows. If you plan on being in the home more than a few years, the investment will pay off for you. With double pane windows the extra pane of glass and the air in between the two panes adds a few extra layers of insulation and therefore comfort to your home. Triple-pane windows are even available on some higher end homes.

Impact Resistant Windows Impact resistant windows are also gaining in popularity with the increasing awareness of hurricanes and the damage they can cause. In fact, in more and more coastal areas, new homes that are in what is called the "wind-borne debris region" are required to have either impact resistant windows, or some other approved window covering or shutters. Shutters are often the prevailing choice of cost conscious builders and home buyers, as impact resistant windows can be very expensive; usually three to four times the cost of standard windows.

Roof Felt

A layer of felt-like material will be placed over the roof decking to provide an extra layer of moisture protection. If shingles get blown off in a storm, water still has the roof sheathing covered by this felt to contend with before it can enter your home.

Shingles

The shingles are now placed on the roof. The most common type of roof shingle on production homes are asphalt shingles because they are relatively inexpensive and they get the job done. Asphalt shingles will be nailed down to the roof. Asphalt shingles come in 5-year to over 50-year ratings. The higher the rating, the more substantial the shingles and thus the higher levels of winds they can withstand.

The builders marketing materials will usually specify the rating of the shingles they install.

You should keep in mind that if you get, for example, a 30-year rated shingle, in all likelihood it will not last you 30 years. Florida's weather fluctuations from warm to sweltering and dry to wet, can take a heavy toll on roof shingles. The factory ratings are for the shingle's lifespan under ideal weather conditions.

Tile and Metal Roofs Alternatives to asphalt shingles are tile, either made of clay or concrete, and metal roofs. Though each is more expensive than asphalt shingles they will both last longer and require a little less maintenance. The covenants in the neighborhood where your home is being built will sometimes dictate what type of roof your home must have, but if given a choice, you might consider a tile or metal roof.

Tile roofs have a long lifespan if installed and maintained properly. Roof tiles are made in a variety of colors to blend with your home's specific design needs. Trusses must be engineered to hold extra weight as tiles weigh considerably more than asphalt shingles. Impact such as that caused by hail can break tiles, and you should have them inspected periodically for damage to prevent problems down the road. When installed properly, roof tiles are less likely than asphalt shingles to blow off in inclement weather.

Metal roofs are also gaining in popularity, partly due to the resurgence of the "Old Florida" architectural look, and also because of their durability. Some roofing companies are offering lifetime warranties for metal roofs, which is a very attractive feature. Metal roofs are also available in different colors to match your home's design. Though the metal material itself reflects the sun's rays, it has a low r-value because it is a conductor of heat, but dead air space and attic insulation can be utilized to increase energy efficiency. As you could probably imagine, noise can sometimes be an issue with rain or hail storms on metal roofs, but sound-deadening insulation can be used to mitigate this problem. Some metal roofs can become dented when hit by falling objects like hail, but some manufacturers offer a "no-dent" guarantee.

Plumbing Top Out

At this stage toilets and bathtubs will be set and the plumbing lines will be stubbed out. Tubs will usually be made of marble, fiberglass, or acrylic and this will be spelled out in the builder's marketing materials. Jetted or "whirlpool" type tubs are usually available, and are especially nice in master bathrooms. Tubs should be covered with either cardboard or plastic during the remainder of the construction process as they can easily be damaged by a dropped hammer or similar accident.

You usually have a choice between elongated or round toilets. Also, "comfort" commodes that are a few inches higher than standard toilets for "ease of use" are usually an option. Check to be sure that the tubs and toilets that are installed are the color and style you have selected.

HVAC Rough

HVAC, which stands for heating, ventilation and air conditioning, will be roughed in at this time. Your air ducts, air vents, and air returns, which regulate the temperature throughout your home, will now be installed. Each room should also have its own air return. This is important because air returns keep your home's air temperature balanced. In homes where this has not been done, you can often have as much as a ten degree difference in temperature from room to room.

Air Conditioner Size Most people when comparing air conditioning units are overly concerned with tonnage, or how big the unit is. But in reality, the tonnage is something that has a point of diminishing returns, meaning that bigger isn't necessarily going to give you better results. The local building codes will dictate the size, or tonnage that the A/C unit used in your home should be, based on the size of your house. The size of what will be installed might also be spelled out in your builder's marketing materials or sales contract.

SEER Rating Here's what is important. Your air conditioning unit will have what is called a SEER rating, or seasonal energy efficiency rating. Many older homes have A/C units with SEER ratings as low as 10. Today the federally mandated standard is 13. The higher the SEER rating, the better the performance (energy efficiency) of the unit. Most builders will let you upgrade the A/C unit so that you can get a higher SEER rating unit if you desire.

Electric Heat Pump In all likelihood your heating system will be an electric heat pump, also not very popular up north. This system will use the same ducts as your air conditioner. Electric heat pumps are not very efficient in temperatures below freezing. At extremely low temperatures an electric heating element kicks in to help the system out. Up north that might be on constantly. But here in Florida, electric heat pumps are the most cost effective and efficient heaters you can have installed in your home. The temperature is rarely below freezing, allowing your heating system to run in its most efficient state most of the time.

Electrical, Phone, Cable and Security System Rough In

Now all of your electrical outlets, cable outlets, phone jacks, and the security system will all be wired in. There are certain minimum requirements for the number and spacing of electrical outlets. Most new homes far exceed these minimum requirements.

Smart Box Located in the garage, closet or some other out of the way place in most new homes is what may be referred to as a smart box. The smart box is the central hub for all cable, phone, and high-speed Internet lines running into and throughout your home. Essential in today's high tech world of networked computers and home electronics systems, with a little instruction you can have multiple computers networked throughout your new home, Internet access in any room you choose, and high speed data flowing to your Tivo. You'd be taking a technological step backwards if you were to buy a new home today without this feature.

Outlet Locations Well before this stage, preferably at the time of selections and before the builder submits for the permit if possible, you should let the builder know if you have any special outlet location requests. Otherwise, it will be a costly mess to have outlets moved or added later, when drywall has to be cut and removed to make any changes.

Think for a while about your living habits now. Do you currently watch television in the kitchen while cooking? If so, request a cable TV outlet in the kitchen. Like to surf the web on your laptop while laying in bed? Make sure there is a high-speed Internet connection near where your bed will go. I think you get the idea. Note that some builders will charge for this service and some will not agree to do it at all, but this is something you should ask about while shopping for a builder.

GFI Outlets Special electrical outlets will be installed in your kitchen, bathrooms, laundry room, garage, and in any other indoor and outdoor locations where water might commonly come in contact with the outlet. These outlets, clearly marked GFI or GFCI for Ground Fault Circuit Interrupter, are safety devices designed to prevent electrocution.

At the first sign of trouble, such as water coming into contact with electricity, they are designed to shut-off or kick the breaker to that particular outlet and prevent you from being electrocuted. Make sure that whoever does your walkthrough with you, shows you how these operate and that they check to make sure that they are functioning properly.

Security Systems One of the most popular options on new homes today is the security system. Whether for a personal residence or a part-time second home, it's nice to have the peace of mind that a security system can offer. The security system is typically comprised of a base unit where the system is controlled from; one or more motion detectors; and sensors placed on doors and windows that alert you if one of them is opened. Other accessories include glass breakage sensors and carbon monoxide sensors. Make sure that all of your home's smoke detectors are hard wired into the system, so that if one of them should detect smoke, you as well as the fire department, will be alerted. An intercom system can be integrated into most security systems, and video surveillance technology is also available at a much higher cost.

Just because your home has a security system, don't sleep easy yet. You are not really protected until you have that system monitored by a security company. Probably the two most popular national security system-monitoring companies are Brinks and ADT. Ask your friends, family, and neighbors for suggestions on companies as well, as there are many other options available from lesser-known local companies. When you contact them they will send a representative out to your home to make suggestions as to the type of monitoring services that will suit you and your system the best. Once you choose the company you wish to go with, you will sign a contract for their monitoring services. When an alarm "event" occurs at your home, whether you are there or not, their monitoring center will be alerted and they will contact the proper authorities.

Alarm.com One other monitoring service that is worth looking into is offered by Alarm.com, from their website of the same name. Alarm.com offers some advantages over traditional monitoring companies, namely wireless monitoring, and an online user interface. Most security systems communicate with the monitoring service via your phone line. Should a burglar cut your phone line, the monitoring center cannot be alerted when they break in. With Alarm.com's wireless monitoring this problem is eliminated as the system uses, you guessed it, wireless signals (similar to how a pager works) to monitor your home.

The online user interface is a convenient feature, especially if this is a second home for you and you are away from it quite a bit. You can keep a virtual eye on your home from anywhere with an Internet connection. From the web you can set the alarm, schedule lights to come on at certain times, and, with available sensors, even monitor things like whether or not the pool guy came to clean the pool last week. You can also be notified of any system event (alarm or door opening, for example) to your Blackberry, cell phone, or by e-mail.

> For more information on security system options that might be available to you, visit: http://www.adt.com, http://www.brinks.com, or http://www.alarm.com

Exterior Doors

The exterior doors, such as the front door and any side exterior doors will now be installed. They can be made of solid wood or a wood composite, metal, or more commonly these days fiberglass. Fiberglass is extremely durable, less likely to dent than steel, and easy to paint.

Framing Punchout

As mentioned earlier, any last corrections that need to be made to the framing before inspection will be done at this point.

Framing, Electrical, Plumbing, and Mechanical Inspections

Next, a whole slew of inspections happen. Typically, you won't even be made aware of it if some aspect of the home fails inspection. But not to worry, the house will not proceed until everything that may be wrong with it has been corrected and re-inspected. Failed inspections happen, and it shouldn't give you the impression that you are getting a substandard home. You should be thrilled that someone has been careful and diligent enough to catch any mistakes, to prevent problems for you down the road.

Stucco

The outside of the home is now stuccoed, providing another layer of protection for your home to the elements. Stucco is basically a muddy mixture of cement, dirt, and water applied to the outside of the block walls of your home. Before the stucco can be applied to the

house, a layer of sheathing and a wire lathe must be placed over any exterior wood framing, such as porch ceilings, to protect the wood and to help the stucco adhere. Blocks walls, however, can have the stucco applied directly to them.

Exterior Paint

Now the outside of the home will be painted the color you selected. Several builders in Florida, after having experienced never before seen water penetration through block walls during the tropical season of 2004, have started using elastomeric paint to help keep wind-blown rain from penetrating the outer shell of their homes.

It is highly recommend that you seek out a home builder using this type of paint, or have your home painted with it soon after you move in. The elastomeric paint is a "waterproofing" paint, not necessarily waterproof, that is applied and a factory representative will usually inspect the application to ensure that it was done properly.

Insulation

Different types of insulation will now be placed in different parts of your home. For example, very thin foil insulation will be placed inside the exterior block walls, batted insulation will be placed between the studs of some interior walls and—if possible—in parts of the ceiling. The remainder of the ceiling will have blown-in insulation, especially in those areas that are hard to reach. Thick, fire rated insulation will also be placed between the garage and the interior of the home to form a fire barrier.

R-Value The effectiveness of the insulation to resist heat from entering the living areas of the home is measured in what is called an R-Value. In Florida, all builders and sellers of new construction homes are required to disclose to you what the R-value is of the insulation installed in the different areas of the home. This will most often be in your sales contract but it could be under a separate "insulation addendum" to the contract.

Garage Insulation Some builders insulate the garage ceiling, and some do not. You should be able to find out from the builder's marketing materials whether or not they do it. If not, and you plan to spend any significant amount of time working in your garage, I suggest you pay a little more to have insulation installed over the garage. It can make a big difference in the comfort of your garage, especially during the hot summer months.

Insulation Inspection

Because insulation is such an important component in your home from both a safety and comfort standpoint, there will be an inspection to make sure the right type and right amount of insulation have been installed properly in the required places.

Soffit and Fascia

The soffit and fascia are aluminum or vinyl materials that are used to cover the eaves, or where the roof overhangs the outside walls of the house. The soffit is designed to prevent water and bugs from entering the home, while still allowing air to flow into and cool the attic.

Drywall

The interior walls of your new home in Florida will most commonly be constructed of drywall, also referred to as "wall board." The drywall will be nailed to the wood studs inside your home and the seams will be hidden by tape. The rooms of your new home are really taking shape now.

Drywall is typically less expensive and less durable than the plaster walls that you may have had or seen in older homes. Drywall is available in different thicknesses, the most commonly used thickness being ½ inch thick. Thicker drywall is generally preferred because each increasing level of thickness adds extra insulating, durability, and sound deadening properties. Also, thicker drywall, especially 5/8 of an inch or thicker, is easier to hang pictures on than ½ inch drywall. Different types of specialized drywall are also available albeit at higher costs such as fire-rated drywall or sound deadening drywall. As with anything else, having the builder install thicker drywall or any specialized drywall will usually result in an increase in costs.

Over the first year or so in your new home, drywall cracks, and nail or screw pops may appear as the house is breaking in and going through a shrinkage process. Near the end of your initial warranty period, usually one year, you should have the builder repair these minor cracks and nail or screw pops.

Windowsills

After the drywall is installed but usually before it is textured, the windowsills will be installed. Most production builders these days are using cultured marble or solid surface windowsills, but some use real wood sills. Many experts prefer cultured marble or solid surface as opposed to wood because windowsills sometimes can get wet if you leave a window cracked open accidentally, and moisture and wood don't go well together.

Drywall Texture

Your drywall job will not be complete without a layer of texture applied to give it some depth, added durability, and also to help hide any imperfections. Two of the most common types of texture being applied in Florida today are called "knockdown" and "orange peel."

Knockdown texture can best be described as looking like splatter, while orange peel looks like—you'll never guess—the peel of an orange. Looking through the builder's model homes, speaking to the sales staff, or reviewing their sales literature will give you an idea of what types of texturing they use.

I briefly mentioned plaster walls before and if you prefer the look and feel of plaster, it can be applied to certain types of drywall. Just check with your builder for the ability to upgrade, and their willingness to do that for you.

Trim carpentry

At this stage a trim carpenter will go into your new home and install the baseboards, special moldings such as crown molding if offered by the builder, chair rails, and doorframes. Interior doors will also be delivered to the home around this time, but won't be installed until after they are painted.

Interior Paint

Your inside painting will be completed at this time. Make sure to request that two coats of paint be applied. Don't fret if the paint does not look perfect at this point. There is still work to be done inside, and a final paint touch up will occur before your home is complete.

Most builders will offer you a choice of colors that you can choose from when you make your initial selections. Be aware though that some production builders do not allow you to choose, and only offer white. If this is the case, you'll either have to paint the interior of the home the colors you want yourself, or hire someone to come in and do it for you.

One tip I can give you is that if you are hiring someone to do the work for you, they will usually quote you a lower price if there is no furniture for them to move or have to cover up. So if possible, have your new home painted before moving anything into the home.

Garage Door

Your garage door will be installed about this time. The garage doors for your home can be made of wood, fiberglass, or most commonly on production homes, steel. Garage doors are rated based on their "wind load" which is the amount of positive and negative pressure they can withstand.

Most people aren't aware that roughly 80% of hurricane damage to a home starts with wind entering through the garage. This makes the garage door the most vulnerable part of your home when it comes to hurricanes, so you'll want to make sure that the garage door on your home is sturdy and made to withstand high winds. This is usually accomplished with the use of heavier, sturdier door materials; door insulation; and many horizontal rows of steel bracing.

Specifications for your garage door were likely submitted with your house plans to the permitting office in order to certify that it is up to local code, and an inspector will verify that the proper door was installed when inspecting your home during construction. In most cases additional bracing can be added, just contact a garage door contractor in the area to come take a look and tell you what options might be available to you.

Tile

Ceramic tile is common in most high quality homes in Florida, at least in "wet" areas such as kitchens, bathrooms and utility rooms. Lower cost homes will substitute linoleum, which can be made to look like real tile. Available in different sizes, usually 12 x 12, 16 x 16, or 18 x 18, tile can be installed pretty much wherever you want, is extremely durable, and can be fairly inexpensive. It can also be very expensive, depending on your tastes and where the tile is made. Imported porcelain tile from Italy will be more expensive than ceramic tile manufactured in the United States, but most untrained eyes won't be able to tell the difference. Another option in higher end homes is travertine marble. In the end, the tile you choose to have put in your home will be a reflection of your personal taste and budget.

A quality tile job will have grout lines of consistent size, and there will not be any high spots or low spots in the tiles. A good tile layer will employ the use of a level to ensure an even application with no high or low spots. Once you move in to your new home it is recommended that you have your grout sealed, or do it yourself, to prevent stains. While the tile can easily be cleaned with water and a mop, grout is much tougher to get clean so it's best to protect it from the start.

Lay out and pour driveway and sidewalks

Now your driveway, front and back patio, as well as any sidewalks, will be laid out and formed up. Once they are formed, they will be inspected to ensure they are the correct size and shape. Assuming they pass inspection, these areas will be poured with concrete.

Final Survey

With all of the components of the home, structure, driveway, sidewalks, and patios in place, a final survey will be conducted to ensure all components are within the setbacks where they are supposed to be.

Irrigation System

Your irrigation system, which can be run off your main water system, a reclaimed water system, or a deep well, will be installed at this point. Be aware that with a well, there is a chance that if the well is not dug deep enough, high levels of sulfur in the water can discolor your exterior paint job over time. Typically, the deeper the well, the less paint discoloring sulfur will be present in the water, and proper depths of deep wells will vary from area to area.

The irrigation system is usually controlled by a timer box that can be manual or electronic, allowing you to set watering times, and setting the system to water certain days, while skipping others. Several Florida cities and counties are experiencing severe water shortages because of rapid growth and development, along with other factors. Therefore, restrictions on how often and when you are allowed to water your lawn are sometimes in place. Check with your city water department for more information on what restrictions might be in place in your area.

Landscaping

Any landscaping that is included with your home will be installed, following the installation of your irrigation system. Most builders include a basic landscaping package with your home, and some will give you the option to upgrade that package. Otherwise, you may want to add some more landscaping after you move in, since most basic packages can be pretty sparse.

Boulevard Trees You might even get some landscaping you don't want or didn't ask for. Many cities and towns are now requiring that developers in new subdivisions plant certain types of trees in the front yards of homes, in the hopes that they will one day add to the beauty of the community. These are typically referred to as shade trees or boulevard trees and are most commonly oaks or magnolias. Cities are also going as far as specifying the number of shrubs that must be placed on the lot.

Measure Cabinets

In the case of custom, or even semi-custom cabinetry, the cabinet company will measure for those shortly after the drywall is in. While some minor corrections to cabinetry can be made in the field, a quality cabinet company will rely on accurate measurements at this stage to build your cabinets to exact specifications.

Cabinets

Cabinets range from entry level laminate cabinets to a step higher with thermofoil cabinets, which are essentially vinyl-covered particleboard, to faux wood cabinet doors with plastic or particle boards drawers and shelves at a step higher, to all wood cabinets at the highest end. Maple, hickory, oak, and cherry are the most popular types of wood cabinets and various stains and glazing such as cinnamon or pecan can be applied to each.

Again, just like most of the other important selections going into your home, the cabinets you choose will be determined by your personal preferences and most definitely budget. Be sure to put a lot of thought into your cabinetry, because if you decide you don't really like it that much a few years down the road, it can be expensive to replace.

One simple way to save a little money without sacrificing design is to have upgraded cabinets installed in your kitchen, and have standard cabinetry installed in your bathrooms. This way, everyone can marvel at your beautiful cabinets in the place where most people tend to congregate, the kitchen.

Counter tops

Your kitchen countertops are one of the most used components of your new home. Your choices range from inexpensive yet functional laminates, to solid surfaces like Corian®, granite, and Silestone®. Again, just like other components in your home, the product you ultimately choose will be based on factors such as personal taste as well as budget. A laminate counter will usually arrive with your cabinets, while solid surface counters must be custom measured for after your cabinets go in. Because of this there is often a couple week delay between your cabinets being installed and the countertop installation.

Laminate Laminate countertops are made in a variety of colors, patterns, and textures. Laminate countertops are among the least expensive countertops available, yet many laminate countertops resemble the look of higher priced counters at first glance. Laminate countertops are not as durable as solid surface counters and can be cut or scorched easily, so you'll need to always make use of a cutting board when working with knives, and hot pads when using hot pots or pans.

Corian® Corian®, invented by DuPont, is one of the most popular countertops available today and with good reason. It is extremely durable, nonporous (this makes it stain resistant), and with over 100 colors available, it can be used in a number of creative applications. Something you might want to think about is that some people decide to transition their solid surface countertops into their bathrooms as well, something you may not typically want to do with laminates.

While Corian® is extremely durable, care should still be used when cutting or cooking. Its cost far exceeds that of laminate counters. A typical kitchen will cost a few thousand dollars as opposed to a few hundred with laminate, but with the proper care and precautions, it should last you a whole lot longer.

Granite Granite is a natural stone, quarried in several locations around the world. Available in a vast range of colors and patterns, granite's use as a countertop surface is very popular, especially in higher-end homes. Because it is a natural stone, no two pieces will ever look exactly the same, allowing the homeowner to express their own sense of flair and uniqueness.

Being that it is a natural stone, granite is porous, and must be sealed regularly to prevent staining. Most experts recommend that this be done twice a year. Granite is known for its hardness and durability, but again, as is the case with other solid surfaces, it is not scratch or burn proof.

Silestone® Silestone® is a nonporous solid surface material made mainly of quartz, the fourth hardest natural mineral. Harder and more durable than granite, Silestone® is scratch, stain, and scorch resistant. However, the manufacturer recommends that as with any other solid surface countertop, proper precautions against each of those be taken. Because it is nonporous, it does not need to be sealed like granite does.

Like each of the other countertop options, Silestone® is available in a variety of colors and patterns, and can be used in a variety of ways. Also, the company that manufactures Silestone® recently introduced Microban®, an antibacterial product built-in to the Silestone®. This helps to continually fight bacteria between cleanings. It should also be noted that previously mentioned DuPont has a similar quartz based countertop product called Zodiaq®.

Appliances are ordered

If you have not picked out your appliances by now, you better get started. At least some of them, like the range, dishwasher, and any built in microwave or wall ovens will be delivered around this time. Your builder will usually have you go to visit their supplier to have you pick them out, otherwise you might be stuck picking out your appliances from a catalog which really makes it tough. Your final appliances—refrigerator, washer and dryer—will usually come a little later.

Appliances might be one of the hardest items to pick out. You should do a lot of looking around at different makes and models before you decide. Appliances are usually available in white, black, bisque, or stainless steel. You should choose a color that coordinates with the rest of your kitchen.

Plumbing Trim

This is when your faucets will be installed. Popular finishes include polished or antique brass, chrome and nickel. You will most likely have picked out the type and style of fixtures you want at your appointment with the builder's decorator.

Mirrors and Shelving Installed

Your bathroom mirrors get installed around this time. Mirrors range from standard flat mirrors, to more upscale beveled mirrors available in a variety of shapes. Some builders offer you a choice at your design meeting and some do not. If not, you can always have them changed out later.

Also, closet shelving, which is usually wire shelving in production homes, will now be installed. Some builders are offering the choice to upgrade to California® type closets, which provide a more useful and appealing shelving and hanging system customized to fit your needs. If your builder does not offer this but you just have to have it, once you move in just look up "closets" in the phone book and get some estimates.

Electrical and HVAC Trim

Your outlet covers and light switches will be installed, along with any light fixtures you have chosen for your home. Also, your air conditioning vents and return vents will be installed if they have not been already.

Attic Insulation

You might recall that when the insulation was installed, certain areas of your home's ceiling may have gotten batted insulation. At this stage the rest of your home's ceilings will get insulated with blown-in insulation. A large tube or hose runs from an insulation truck and the installer climbs into the attic and "blows-in" the rest of the insulation. They will usually apply it until there is at least a foot or more of insulation covering all areas.

Flooring

Any hardwood flooring or carpet you have requested will be installed at this time. You will have picked out your flooring choices at your meeting with the decorator, and now you get to see how it finally looks installed in your new home. If you're not excited yet, you should be. You're almost home!

Final Inspection

This is the last city inspection that your home will have. After it is complete, and assuming your home has passed, the city will then issue a Certificate of Occupancy (or "C.O." in builderspeak), basically stating that the home is habitable and has been inspected to meet or exceed local building standards.

Paint Touch-Up

The painter will go back in and touch up any spots he might have missed, or areas that have been scuffed, nicked, or otherwise damaged as things have been delivered and installed and people have been working.

Meter Installed

Up until this point, your home has been running on temporary power. Now the local power company will come out, remove the temporary power pole, and install an electric meter. A power company employee will check this meter each month to see how much power has been used, and the power company will use this information to calculate your bill.

Hot Check

With the home running on permanent power, all of the electrical functions of the home will be tested to make sure there are no shorts in the system, and make sure everything functions the way it should.

Install the rest of the appliances

With 99 percent of the work on your home complete, your final appliances will usually be delivered and installed. You will have a chance very soon at your final walkthrough to verify that these are in no way damaged, and that they work like they should.

Punchout

Here the last one percent of work, which is often the most important, will be attended to. Either the superintendent or a walk-through specialist will walk through your home examining it for quality finish and attention to detail. Anything out of order or not in tip-top shape will be taken care of.

Final Cleaning

After all construction work is done, a cleaning crew will go through the home cleaning it from top to bottom, mopping floors, vacuuming carpet, scrubbing toilets, and cleaning counters. After the cleaning it may be found that certain counters or tubs have been scratched during construction. If this is the case they will be buffed or otherwise repaired.

14.
YOUR NEW HOME WALKTHROUGH

The walkthrough, or new home orientation as it is sometimes called, is one of the most important phases in the construction of your new home. It is a time for you to meet with the builder and let him or his representatives acquaint you with your new home and all of its components. The walkthrough is also a time for you to give your new home the once over, looking for any construction issues not up to quality standards. Here is what could be considered the ideal walkthrough in detail.

ALLOW ENOUGH TIME

Allow ample time to go through your new home. In my experience an hour and a half to two hours is sufficient for average sized new homes. Also, leave any pets, kids, or curious friends and relatives at home. There will be plenty of time for them to experience and enjoy your new home in due time. The walkthrough is serious business and should be treated as such. Minimizing distractions is critical.

WHAT TO BRING

To ensure a successful walkthrough bring along several pens or pencils, a black permanent marker, a packet of neon green dots available at office supply stores, a pad of legal paper, some bottled water, and a ton of patience. Understand that everything might not be perfect once you start the walkthrough. It's just the nature of home building that no matter how careful, the builder can't catch everything. But, if you follow my advice, the builder and his employees will be in the position to get things corrected for you in a timely fashion.

The order of the walkthrough is not really important as long as everything gets covered. As you find items not up to standards, place one of the neon green stickers I suggested you bring on the item and write it down on your legal pad or a punchlist provided by the builder, or both if you feel it necessary. Green dots can mysteriously disappear but if you write it down it can't be forgotten for long.

BREAKER BOX AND ELECTRICAL SYSTEM

You will of course be tempted to head for the front door and bask in the glow of your fresh new home. But not so fast. Let's cover some things in the garage first. The garage houses several important components of your new home and you should become familiar with them. The first item on the list is the breaker box. This is where the electricity that comes into your home is regulated. The walkthrough representative should show you where it is and how to operate it.

Make sure that each breaker has been clearly labeled for you. This will eliminate headaches down the road. Also, there should be some GFI outlets in the garage. Now is a great time for the walkthrough representative to test those in front of you, and to show you how they work. Also, make sure they test the GFI outlets inside the home when you get in there.

HOT WATER HEATER

Be sure to check the hot water heater. Make sure the size, measured in gallons, is what you contracted for. The walkthrough representative should show you how to turn it off so you will know how to when necessary. There are timers available for your hot water heater that can easily be installed that will save you some money on your electric bills. If your hot water heater comes with a timer, have the walkthrough representative show you how to set it.

WATER SHUTOFF

The main water shutoff valve to the home will usually be located inside the garage or sometimes on the outside. The walkthrough representative may advise you to turn the water off if you will be leaving the home for days at a time. This is probably good advice, at least initially until you've lived in the home a while and made certain there are no leaky toilets or pipes.

If you do turn off your water, make sure that you also turn off the breaker for the hot water heater. The hot water heater has coils inside that can burn up if there is no water passing through. When you return home, it is very IMPORTANT to make sure you turn the water back on before turning the hot water heater back on.

AIR HANDLER AND AIR FILTER

The air handler, which distributes the heated or cooled air throughout your home, will usually be in the garage as well. Make sure the walkthrough representative opens the filter door to show you how to change the air filter. Using the black permanent marker, make note of the filter size in a conspicuous place on the front of the air handler. You should change the air filter about every month for best performance.

GARAGE DOOR

While you're still in the garage, open and close the garage door to check for proper operation and make sure the remote controls work. If your garage door opener came with an outside keypad, ensure that it too works. In the event of a power outage you may need to open the garage door manually. Have the walkthrough representative show you how to do that.

KITCHEN

Once inside the home, the best place to usually start is the kitchen because there is so much to cover there. Make sure that there are no scratches on the kitchen countertops or cabinets. Open and close a random selection of cabinet doors to make sure they are working properly. Make sure the hinges are tight, and the cabinets aren't sticking or rubbing against anything as you are opening and closing them. The representative should give you care and cleaning instructions for both your counters and your cabinets.

Turn on the kitchen faucet and set it to the hottest setting. Here we are checking to make sure that the hot water heater is working properly. As long as you've got hot water after what you feel is a reasonable length of time, you're doing just fine. Have the walk-through representative show you how the sink disposal works, and how to clear it if it gets clogged. Also have them show you where the individual shutoff valve is for the water in the kitchen as well as the locations of the GFI outlets.

APPLIANCES

Examine the appliances that came with your home. First, examine the outside of them to make sure there are no scratches or dents. Accidents do happen during construction, but assuming you bought new appliances, and not scratch-and-dent specials, they should be in brand new condition. Turn the stovetop on, check that the burners are working, and then try heating the oven. Assuming everything is working thus far, start the dishwasher to run through a cycle. This is to mainly make sure that there are no leaks in the dishwasher, either when it fills or when it drains.

While the dishwasher is running do a quick check of the refrigerator. If there are integrated ice and or water controls in your refrigerator make sure they work. Don't use the first batch or two of ice; just discard it in the sink. Also, most manufacturers suggest running through and pouring out the first couple of gallons of water from the refrigerator. This is to make sure that the water line becomes clear of any debris that may have gotten inside during construction and installation.

If your home came with a microwave, also check to make sure it works. In the laundry room, start both the washer and the dryer if provided and make sure they are working correctly. Make sure the dryer vent hose is connected.

All of the appliance instructions and warranty information should be kept in one easy-to-access location. Some of them may have cards for you to fill out and mail in to the manufacturer to record your warranty. These cards are primarily used to collect marketing information, and you don't really have to fill them out to get the warranty.

DRYWALL AND FLOORING

Before leaving the kitchen, examine the flooring for quality. Also check the walls for any drywall imperfections and check the paint for any spots the painter may have missed. As you see things that don't meet your standards, write them down on the list and place a green dot on or near the problem area. This is so that the drywallers or painters know exactly where to look to correct the problem areas.

Continue your flooring and wall inspection throughout the remainder of the home. Don't forget to look up every now and then and inspect the ceilings.

SYSTEMS AND COMPONENTS

As you are going through the home, have your representative show you how various things work, such as how to set and control the thermostat, how to use the security system and intercom if there is one, and how to operate the central vacuum if you bought one. If your home has a fireplace, whether it is wood burning, gas, or electric, have the walkthrough representative show you how it works. Make sure you are given instruction booklets on each of these items and that you place them with your appliance booklets.

BATHROOMS

Visit the bathrooms and check that the plumbing works. Again turn on the hot water, then the cold water to check the functioning of each. Be sure to check the showers and baths, as well as the sink. Water lines sometimes get reversed. Hot will be cold, and cold will be hot, but this can be easily corrected. Flush the toilets and make sure they have adequate water flow and don't remain running long after you flush. Check the tile work inside the showers to make sure that there are no holes or gaps in the grout or caulking. You don't want water getting behind your tile in there. Examine the vanity tops for scratches and cabinets for loose hinges.

EXTERIOR

Be sure to inspect the outside of your home as well. The walkthrough representative should familiarize you with where the hose bibs are located, the sewer cleanout, the A/C unit and anything else that is important. Make sure all of the exterior walls of the home are evenly painted, and do an inspection from ground level of the roof to make sure there are no shingles that look loose or out of place. If your home comes with a sprinkler system, you should be shown how to operate that.

WARRANTY

After you feel you've examined the home top to bottom and have made note of anything that is not satisfactory, you should have the walkthrough representative go over any warranty paperwork that is given to you, so you have an understanding of what items in the home are covered and for how long. Most warranty plans cover most everything for a short period of time, usually the first year. The systems of the home, things like plumbing, electrical, and HVAC, will be covered for a little bit longer, maybe up to two or three years.

There will also be a warranty on the structure. This is the longest lasting component of the warranty. When you hear a builder say a ten-year warranty or 15-year warranty, they are referring to the warranty on the structure. The structure is usually deemed to include the foundation and footings, beams, lintels, columns, walls, roof framing systems and flooring systems.

When things settle down a little bit and you have some time, it can never hurt to read over all of the warranty information. This will help you feel more comfortable with the warranty claim and repair process should you ever need to go through it in the future.

EMERGENCY INFORMATION

The walkthrough representative will usually give you a list of subcontractors who worked on your home so you can call them if you have a problem with something. You should also be sure that you have a list of repair people to contact should an emergency arise on a weekend or during any non-business hours.

These people should include the heating and A/C contractor should the heat or air break; the electrical contractor if you lose power due to something other than a loss of overall power from the power company; the plumber for if your hot water heater breaks or if there is a sewer stoppage; and finally the number for the roofer if you get a roof leak. I also recommend having the number for a

24-hour water extraction company handy, just in case a pipe breaks or a water heater bursts and your home is flooded.

Write all of these numbers down on one piece of paper and tape them to the inside of a cabinet so that you can find them easily in an emergency.

SIGN HERE PLEASE

To conclude the walkthrough, the walkthrough representative will typically have paperwork for you to sign stating that he walked you through and familiarized you with everything in the home, and that all the workmanship was satisfactory. Just make sure that the items you found to be unsatisfactory are either on this paperwork or will be attached to it in some form or fashion.

It is not absolutely critical that these items all be completed before your closing, so long as they are documented as needing repair. Invariably in the days and weeks after you move in, you will find more items needing the builder's attention. Just write all these items down as you find them and bring them to the builder's attention.

It has been a long process but now you are all set to enjoy what you have longed for, a beautiful new home in Florida.

15.
SWIMMING POOL CONSTRUCTION

Some of you buying a home in Florida will have a swimming pool built along with your new home, or put in once your home is finished. Here is how the process of building your pool will generally go.

First you will meet with the pool builder to go over all that you want in your pool. Visit several pool websites, buy some pool magazines and talk with other pool owners to get an idea of what you want. You'll need to decide on the size, the shape, the depth, where steps and ledges will go, waterfalls, spas, heaters, special jets, tile, interior coating (marcite) colors, and more.

The builder will submit plans for your pool to the city and wait to get the approved permit back before beginning. Once approved, the pool builder will lay out your pool according to the shape you requested and the city approved.

A backhoe will be used to dig out the pool, a truck will haul off some of the dirt, and some will be left to backfill and grade the pool deck later. Steel rebar will be placed in the pool to form a sort of "basket" in the shape of your pool.

At this point, drains, jets, and other plumbing will be installed. The city will inspect that the steel and pre-plumbing have been done up to code.

SWIMMING POOL CONSTRUCTION

Then, fast drying, high strength concrete will be applied to create the shell of your pool. This process is called "shooting" the pool because the concrete is shot out of a hose. It actually looks like a pool now.

The excess dirt from digging the pool will be used to fill in around the outside of the shell and to grade the deck. Tile that you chose will be installed around the perimeter of the pool at this time.

Electricity is run to the pool equipment location and the equipment may also be installed around this time. The electrical work will be inspected as well as the plumbing lines. Assuming all is working well, the pool deck will be poured. Once the concrete dries it will be textured and coated with an acrylic type coating, which should last you several years with proper care.

Your screen enclosure, which usually requires a separate permit to ensure proper engineering and strength, will be done at this time. Now, the interior of your pool will be cleaned out and acid washed. It has probably had standing water, dirt, and construction debris in it, but no longer. The interior pool surface, most commonly a material called marcite is applied.

Your pool is filled almost immediately afterwards. Once filled, your pool builder will start-up and clean the pool and add the proper chemicals.

After you've been given your "new pool owner" orientation, you are ready to enjoy your pool. Some people prefer to maintain the chemicals in and clean their own pools, while others prefer to hire someone to take care of it.

My advice is to leave it to an expert, especially at the beginning. During the first few months of operation your pool goes through a "breaking-in" period and an expert can best control and guide this process. Once all the chemicals and the "mood" of your pool stabilize, you can take over the duties if you wish.

16.
REAL ESTATE CLOSINGS

There are basically two places where real estate closings take place in Florida. The first and most common place is at a title insurance company, and the second is at an attorney's office.

TITLE INSURANCE COMPANIES

Because of their importance in your real estate transaction, most title insurance companies provide closing services. It is very common for closings on property in Florida to take place at a title insurance company office. The title insurance company will act as a neutral third party to ensure that all terms of the contract have been met, and they will collect and disburse funds according to the terms of the contract.

What is Title Insurance?

Before your closing, a title insurance company will conduct extensive research into public records, surveys, and other recorded documents to ensure that no party (other than the seller) holds an interest in or has a lien upon the property you are trying to purchase. According to the American Land Title Association, nearly one-third of all title searches reveal a problem with the title. Unknown heirs, divorces, tax liens, and fraud or forgery can cause title problems. Thankfully most can be resolved before your closing.

Upon completion of their research, the title insurance company issues an owner's policy to the buyer, and a lender's policy to the lender. The seller will customarily pay for an owner's policy but, as with most other costs, this is negotiable. If you do end up having to pay for it, the cost will be $5.75 per $1,000 up to $100,000, and $5.00 per $1,000 thereafter. If you are financing, your lender will require you to pay for the lender's policy, but this does not cost very much. It will be usually be only a couple hundred dollars or less because you are getting what's called a simultaneous reissue credit with the owner's policy.

ATTORNEYS' OFFICES

Many states require that an attorney conduct the real estate closing, but that's not the case here in Florida. Even so, many people prefer the peace of mind that an attorney can bring to a real estate transaction. If your closing is being held at an attorney's office, it will most likely be the seller's attorney, since they usually pay for the title insurance policy. If that is the case, make sure that your attorney gets a copy of what you will be signing and has a chance to review it before you go to the closing. It is not critical for your attorney to attend the closing. Often times your real estate agent will attend as a courtesy to you, just be sure to ask.

CLOSING COSTS

There are costs other than the sales price that are incurred in every real estate transaction. As the buyer, your share of these costs will typically range from 1% to 2% of the sales price. The closing costs that you pay will be a function of a couple factors including what you have negotiated in the real estate contract and whether or not you are getting a mortgage. Costs that are customarily paid for by the buyer include:

- Recording of the deed
- Documentary stamps on the deed*
  ($.70 for every $100 of the sales price)
  *This is typically paid by the seller in a resale transaction,

but I included it here because many builders require the buyer to pay this
- Documentary stamps on the mortgage ($.35 for every $100 financed)
- Intangible tax on the mortgage ($.002 times the mortgage amount)
- Lender's title insurance policy
- Taxes
- Prepaid interest
- Prepaid HOA Dues, capital contributions, or transfer fees
- One year of insurance in full
- Appraisal Fee
- Underwriting Fee
- Flood certification fee
- And more...

HOW SHOULD YOU TAKE TITLE?

How you take title to your home is an important question that should only be answered after consulting an attorney, and your accountant, due to tax and estate-planning implications. How you take title establishes proof of ownership of a property, and often determines who will inherit your home when you die.

Sole Owner

Also known as owner in severalty, sole ownership is how you might take title if you are single, divorced, or widowed. Should you happen to die while owning the property, the property will be passed on according to your will. If you do not have a will, it will be passed on by descent to your heirs.

Tenancy by the Entireties

Tenancy by the Entireties is the most common way to take title in Florida for those who are married. In this case, if one spouse dies, the remaining spouse automatically becomes the owner.

Tenancy in Common

If you are in a second (or third, or fourth) marriage and you have children from a previous marriage, you may want to consider a tenancy in common. In this case, if you should die, your share of ownership in the property can be willed to your children.

Joint Tenants with Right of Survivorship

If you and the person you live with are not married, but you want to ensure that they receive full ownership of the property when you die, then joint tenancy with right of survivorship may be for you. The surviving owner will own the property outright, and the heirs of the deceased will have no claim to the property.

Taking Title in a Trust

Another option many boomers may want to consider is to take title in a trust. Speak to an estate planning attorney or a real estate attorney as to how and why you might take title in this way.

17.
MAINTAINING YOUR NEW HOME

There are several things you should do after you move in to keep your home looking and operating like new.

PERIODIC MAINTENANCE

These include monthly maintenance like changing your air filter, cleaning your disposal blades by running ice cubes through it, and checking for leaks around toilets and under sinks. Periodically check all grout and caulking for any cracks, as this can occur due to shrinkage. Of course, you'll want to maintain your appliances and have them repaired at the first sign of trouble.

Twice a year you should have your roof (please don't try to do this yourself) and the exterior of your home inspected. Check the operation of all windows and shutters if you have them. You should have your heating and air conditioning system inspected and serviced twice a year as well, once at the start of summer, and again before the winter heating season.

Your home warranty information provided by your builder may have additional hints and tips for keeping your home in tip-top shape year round.

PEST CONTROL

In Florida, having a pest control service company come to your home regularly is essential in keeping your home both comfortable and free of insects and other pests. Without it you may have some unwanted roommates. Most non-native Floridians, especially Northerners are surprised at the amount of pests in Florida, but unfortunately the climate helps them thrive. The great weather might be what brought you here, and the bugs like it just as well.

The pest control company will usually start you off with an initial "kill everything in sight" type of service and maintain regular visits every 60 to 90 days. If you are still seeing bugs in your home, most companies will come out in between scheduled visits to try and get rid of the problem. Look in the phone book under "pest control" for companies providing service in your area.

If you are having a new home built, another pest control option is to have pest tubes installed in your walls. This is done after the framing stage, and before the insulation and drywall are put in. In the future, when the pest control company comes to your home, they service the tubes inside your walls from a base station outside, eliminating the need for you to be home when they come and eliminating the need for them to come inside. These tubes aren't something normally offered by most builders, so you might have to ask. They should not be very expensive, because pest control companies that work with them will install them for next to nothing in the hopes of making up for the expense when they sign you to a service contract. Not all pest control companies service these pest tubes, but more and more are doing so as they gain popularity.

18.
FLORIDA RESOURCES

POPULATION AND GROWTH

Florida has 67 counties, fifteen of which are among the fastest growing in the nation. This statistic includes the nation's fastest growing county, Flagler County, located on Florida's east coast, between Jacksonville and Daytona Beach. As a result of the growth, Florida's population has almost doubled, from roughly 9 million in 1980, to close to 18 million in 2006.

Florida is currently the fourth most populated state behind California, Texas, and New York, and Florida is projected to move ahead of New York by 2011.

Florida's rapid population growth is due, in part, to the thousands of baby boomers moving into the state each year. At the time of this writing in 2006, experts project that Florida's population is growing by almost 370,000 people per year. This works out to more than 1,010 people a day.

These statistics might make you want to stop and say hmmm...maybe Florida will be too crowded for me when I am ready to move there. Luckily, this is not likely to happen soon. Florida's local governments and developers are now well aware of Florida's growth rate, and are rapidly, if not frantically, preparing the state with extra infrastructure, housing developments, and resources necessary to accommodate the large influx of residents.

One of the best ways to mentally grasp the geographical distribution of Florida's population is to look at Florida's most populated areas. Distributed around the coastal areas and Central Florida's east-west corridor, almost three quarters of the state's population live in and around 10 major areas.

FLORIDA'S TOP TEN METROPOLITAN STATISTICAL AREAS

1. Miami/Fort Lauderdale/Miami Beach 5,428,962
2. Tampa/St. Petersburg/Clearwater 2,636,972
3. Orlando/Kissimmee 1,953,354
4. Jacksonville 1,277,763
5. Sarasota/Bradenton/Venice 672,231
6. Cape Coral/Fort Myers 549,442
7. Lakeland 541,840
8. Palm Bay/Melbourne/Titusville 531,970
9. Deltona/Daytona Beach/Ormond Beach 494,649
10. Pensacola 440,066

HURRICANES

In Florida, hurricane season is a fact of life. The season officially begins June 1 and runs through November 30, with most activity typically occurring in late August through September. For many people in Florida, hurricanes are an excuse for a party. Most everyone who has lived in Florida for longer than a few years has been through at least one hurricane and several tropical storms. Though they are nothing to dismiss or joke about, with proper planning and precautions you'll be able to weather the storms like a native.

Tropical depressions, tropical storms and hurricanes, when present, dominate the weathercasts here in Florida. The attention is for good reason. Hurricanes produce extreme winds, tornadoes, torrential rain, and storm surge, which can cause severe flooding of coastal areas. In fact, nearly 60 percent of hurricane fatalities occur as a result of flooding. Other effects are of course property damage, power outages, temporary loss of public services, bridge and road closures, loss of communications, and hospital closures.

A hurricane usually begins as a tropical wave that develops into a low-pressure system known as a tropical depression. Tropical depressions are not very organized but they have the potential to become stronger and evolve into more organized storms. You can, and if you move to Florida most likely you will, track storm developments online at sites like weatherunderground.com or nhc.noaa.gov.

Tropical depressions have sustained winds of up to 38 mph. Should the storm gain strength, the next level is called a tropical storm, with winds between 39 and 73 mph, strong enough to cause pretty severe damage to older, unprotected structures.

Hurricane Watch versus Hurricane Warning

- A Hurricane Watch indicates the possibility that you could experience hurricane conditions within 36 hours. This watch should trigger your family's disaster plan, and protective measures should be initiated.
- A Hurricane Warning indicates that sustained winds of at least 74 mph are expected within 24 hours or less. Once this warning has been issued, your family should be in the process of completing protective actions and deciding the safest location to be during the storm.

Hurricanes are rated on a scale of 1-5 on the Saffir-Simpson Scale. Here are some defining characteristics of the five categories of hurricanes, and recent Florida sightings of each:

Category One Hurricane

Winds 74-95 mph.

Category Two Hurricane

Winds 96-110 mph.
Recent Category two hurricane making landfall in Florida: Frances, 2004.

Category Three Hurricane

Winds 111-130 mph.
Recent category three hurricane making landfall in Florida: Jeanne, 2004.

Category Four Hurricane

Winds 131-155.
Recent category four hurricane making
landfall in Florida: Charley, 2004.

Category Five Hurricane

Winds greater than 155 mph
Recent category five hurricane making
landfall in Florida: Andrew, 1992.

The category ratings are based on the strength of the hurricane, not an expected level of its potential for destruction. Hypothetically, a category one hurricane hitting a highly populated area could cause more destruction than a category five hitting a less populated place. To read more about hurricane categories and the Saffir-Simpson scale visit: http://www.nhc.noaa.gov/aboutsshs.shtml

If you do become a property owner in Florida, you should take great care in preparing your property and your household for the threat of hurricanes. Don't assume that because you live in the center of the state that you are immune from storm fallout. Be prepared. Have a written hurricane plan. At the beginning of hurricane season, check your disaster kit, batteries and non-perishable food supply. Tips for developing a plan and building a hurricane kit can be found on-line at redcross.org, flash.org, or similar websites. Local television news stations usually put out hurricane tracking guides, which have tips on what you should do to prepare. These are usually available in local grocery stores.

Also, no matter how secure you may feel in your home, if you are asked by local authorities to evacuate, you should do so and do so early. Prepare for traffic congestion and long lines at the gas pumps. It helps to have a plan in place regarding where you will go and what you will take with you. Yes, it is a pain and inconvenience to pack up your essentials, secure your home, and drive to safe territory, but it could also save your life.

In Chapter 13, New Home Construction, you'll find tips for strengthening your new home in Florida to better withstand the destructive forces of hurricanes. If you plan on buying a resale home not built within the last two years, visit the Federal Alliance for Safe Homes at http://www.flash.org or My Safe Florida Home at http://www.mysafefloridahome.com for tips on how to prepare your home to withstand a hurricane.

GETTING AROUND FLORIDA

Florida is a lot bigger than most people think. The distance top to bottom is 447 miles, and 361 miles side to side. Florida is the twenty-second largest state, with 58,560 square miles. Florida's longest river is the St. John's River, which is 273 miles long. Lake Okeechobee is its biggest lake at 700 square miles, making it the second largest freshwater lake in the continental United States, ranking just behind Lake Superior.

Air Travel

Getting to Florida by air is pretty easy, no matter where you are going in the state. Florida currently operates 13 international airports, with Orlando International Airport being the busiest with more than 34 million passengers a year, followed closely by Miami International Airport with approximately 31 million passengers a year. Florida also has several regional airports, and many smaller executive and community airports which can be a good choice for avoiding long lines and big crowds at larger airports, assuming you can get flights to where you need to go.

Florida airports are serviced by all the major airlines as well as discount and charter airlines. Search your favorite travel site to see how convenient and (usually) inexpensive it is for you to fly to Florida. If you plan to keep a second home here, it should be convenient for you to fly in and fly right back out without much trouble, cost, or inconvenience.

Florida's International Airports (with Airport Codes)

Daytona Beach International Airport (DAB)
Fort Lauderdale/Hollywood International Airport (FLL)
Jacksonville International Airport (JAX)
Key West International Airport (EYW)
Melbourne International Airport (MLB)
Miami International Airport (MIA)
Orlando International Airport (MCO)
Palm Beach International Airport (PBI)
Panama City International Airport (PFN)
Sarasota-Bradenton International Airport (SRQ)
Southwest Florida International Airport (in Fort Myers) (RSW)
St. Petersburg/Clearwater International Airport (PIE)
Tampa International Airport (TPA)

Major Interstates

Florida's major interstate highways are I-95, I-75, I-4, and I-10. They primarily connect the state north to south and east to west.

I-95, which runs up and down the east coast of the United States, reaches Florida north of Jacksonville, and runs the length of the east coast of the state, ending in Miami. It meets with I-10 in Jacksonville, and I-4 in Daytona Beach.

I-75 connects the west coast of Florida with the Midwest states, making cities on the west coast, including Tampa, Sarasota, Bradenton and Naples, hotbeds for Midwest vacationers, second homeowners, and retirees. I-75 begins in Florida about 45 miles west of Jacksonville. Once it snakes its way to the west coast of the state, it passes through Tampa (where it meets up with I-4), Bradenton, Fort Myers, and finally Naples. Here it begins to run east-west to Fort Lauderdale.

I-4 runs through the central part of the state, connecting Tampa, on the west coast, with Daytona Beach on the east coast. Interestingly, in 2004 Hurricane Charley became known as the "I-4 Hurricane" because it entered Florida near Tampa, and followed the path of I-4 very closely before exiting the state just north of Daytona Beach.

I-10 runs the width of the Panhandle of Florida and connects Pensacola with Jacksonville. If you were to keep driving west out of Florida on I-10 you would eventually end up in the Pacific Ocean near Santa Monica, California.

Road Construction

In a continuing effort to prepare the state's roadways and interstates for Florida's growth, road construction projects are always underway. If you'd like to find out where they are located before making a trip, visit the Florida Department of Transportation online at http://www.dot.state.fl.us. Here you can also find other valuable information such as the mileage between major cities and the locations of Florida's rest areas, toll roads, and speed limits.

Speeding

Don't speed. The Florida Highway Patrol and other law enforcement agencies are the butt of jokes such as: "Welcome to Florida, may I have your license and registration please." Be aware that speeding fines are doubled in work zones in Florida. Speeding might leave you with a couple hundred bucks less to spend on your new home here.

511 Traffic Info

Florida offers an in-state travel information system that you can access by dialing 511 from a cell phone or landline. From 511, you can get updates on traffic for major roadways and interstates, as well as construction information, lane closures, and special alerts.

The system is also available online at http://www.FL511.com

Florida Driver's License

If you'll be moving to Florida full time, you'll want to get a Florida driver's license. Part-time Florida residents can also keep their out-of-state license and get a Florida license that states "Valid in Florida Only." This could be handy for situations where you need to provide proof of residency (like to get discount theme park tickets).

Either way, you can get that process started at the Florida Department of Highway Safety and Motor Vehicles website at http://www.hsmv.state.fl.us

Save time getting your license by taking advantage of online appointment scheduling. After you get a Florida license, in some cases, you can even renew online or by telephone.

Driving Miss Daisy

Some boomers moving to Florida will be bringing their elderly parents to live with them, many of whom will continue to drive in their seventies, eighties and nineties. The state of Florida offers a great resource covering many aspects of elderly driving at http://www.floridagranddriver.com

Distances Between Major Cities

To find the mileage between other Florida cities visit: http://www3.dot.state.fl.us/mileage/

DISTANCES BETWEEN MAJOR CITIES

	Daytona Beach	Fort Lauderdale	Jacksonville	Lakeland	Melbourne	Miami	Orlando	Sarasota	Tampa	West Palm Beach
Daytona Beach	x	230	89	107	86	256	54	184	138	189
Ft. Lauderdale	230	x	317	209	145	25	206	202	233	42
Jacksonville	89	317	x	178	173	342	134	240	190	276
Lakeland	107	209	178	x	95	225	54	79	33	169
Melbourne	86	145	173	95	x	170	66	172	126	103
Miami	256	25	342	225	170	x	231	214	249	68
Orlando	54	206	134	54	66	231	x	131	85	166
Sarasota	184	202	240	79	172	214	131	x	57	186
Tampa	138	233	190	33	126	249	85	57	x	192
West Palm Beach	189	42	276	169	103	68	166	186	192	x

EMPLOYMENT

If you plan to continue working after you move to Florida, or if you wish to start a second career, opportunities abound in Florida. The state's job market grew in 2006 and the jobless rate in Florida was the third lowest in the nation. Jobs in construction and professional and business services are growing the fastest. The tourism industry also continues to be a strong economic force, employing almost a million people in 2005.

When looking for work in Florida, start with where you work now. If you like your current job, ask your manager if there might be an opportunity for you to continue working there by telecommuting. With so many boomers retiring, many companies are starting to see their company-wide levels of knowledge and experience start to dwindle. In order to keep that from happening, your company may want to keep you on board, even if you are hundreds of miles away (ON A BEACH!) in Florida.

Looking for something new? If it's part time work you are after, check the classifieds in the local paper near where you are moving. Both local and national companies with locations in Florida actively recruit older employees. Part-timers and those looking for temporary or seasonal work can look to companies like Kelly Services, Manpower, and Spherion.

Other boomers may prefer to work year-round. Opportunities in Florida include retail, health care, communications, finance, insurance, business, and marketing.

In addition to looking for opportunities in local papers, don't forget to search online jobs portals such as Monster.com and hotjobs.com for opportunities that might be available. Also, aginghipsters.com and careerbuilder.com have partnered up to create an online job portal geared specifically to baby boomers. Visit aginghipsters.com and click on "Careers" for more information.

This is your new life. What you choose to do with your time is completely up to you. Just make sure you fill it with things you enjoy.

COMMUNITY SERVICE

April has been proclaimed Florida Volunteer Month by the state cabinet, but don't wait until then to do something for the community. The volunteer rate for baby boomers is the highest of any age group. Invariably when you are ready to move to Florida, it will not be your parent's retirement with time spent solely on the golf course, napping in a hammock, or playing bingo everyday. You are likely to want to spend at least part of your time volunteering for a worthy cause. Luckily, there are plenty of opportunities for you to get involved in community service in Florida. Whether you choose to help out at a local school part time, join the local chapter of the Red Cross or United Way, or help build homes for Habitat for Humanity, your time and experience will be highly valued and appreciated.

As you probably already know, you'll feel great about yourself for doing it. Getting involved in volunteer work naturally increases your quality of life. Many volunteer agencies actively seek out older volunteers for their expertise and availability. Some even organize special task forces made up of their members who are 50 or 65 years old or more.

> For more information on volunteer opportunities for baby boomers visit http://www.getinvolved.gov and visit http://www.volunteerflorida.org for more information on the Governor's Commission on Volunteerism and Community Service.

CONTINUING EDUCATION

One way to fill your time is to take some classes at a local college or university. Classes range from foreign language classes, to cooking classes, to photography classes and more. Maybe you want to learn to speak Italian before a big trip to Florence, or learn the intricacies of Thai cuisine to impress your friends and family. With 11 state universities and nearly 30 independent colleges and universities, whatever you would like to learn, there's bound to be a school nearby that offers it.

Don't be intimidated in thinking that these classes will be full of late teen and twenty-somethings. Boomer participation in continuing education, especially when it comes to subjects that support their hobbies, is rapidly growing. As well, many schools have organizations for students of non-traditional age. There's bound to be a friend or two to make, and a good time had, no matter what type of subject you choose to study.

Also, be sure to take advantage of telecommuting opportunities offered from many institutions. The classes can either be self-paced or have assignment deadlines posted by the instructors. Often, students interact with teachers via email and with other students on class bulletin boards. Like traditional courses, online education runs the gamut from mathematics to art.

Museums and private organizations also offer classes. You can learn about Florida history, butterfly gardening, or surf fishing, explore yoga, take up sailing, or join the crowd at Bike Week by taking a motorcycle safety course.

> For more information on Florida's colleges and universities as well as links to each institution visit http://www.fldcu.org or http://www.fldoe.org

HEALTH CARE

Close to 20 percent of Florida's population is over age 65. This is the highest percentage of all states, and almost 50 percent higher than the U.S. average. Consequently, there are more than 200 community hospitals in Florida, some of which are regarded as being among the best in the nation. You can be assured that there are cutting edge facilities close by, no matter which part of the state you move to.

Florida does a great job of supplying its residents with the information they want to know. Nowhere, perhaps, is this information more important than in the field of healthcare. Be sure to check out the following websites.

MyFloridaRx.com

Prescription drugs can account for a large portion of your health care bills, especially if you don't have insurance, aren't old enough for Medicare, and don't qualify for Medicaid. This site, developed by the Florida Attorney General and the Agency for Health Care Administration, lets you compare prices for the 100 most commonly used prescription drugs. MyFloridaRx.com also lets you see which pharmacy in your area has the best drug prices.

> http://www.myfloridarx.com

FloridaHealthStat.com and FloridaCompareCare.gov

Two more sites dedicated to providing information to Florida's health care consumers are FloridaHealthStat.com and FloridaCompareCare.

gov. These sites go several steps beyond comparing prescription drug prices. At these two sites you can compare hospitals, pharmacies, physicians, nursing homes and even health care plans, based on different criteria relevant to each.

At FloridaHealthStat.com your search options include prescription drug assistance programs, physicians by specialty, HMOs and insurance and health facilities. Also, if you have elderly parents who may need a nursing home soon, check the Nursing Home Watch List at FloridaHealthStat.com before sending them.

At FloridaCompareCare.gov you can view hospital and care center profiles, which include the types of care available at each location and facility mortality rates. This site also has a facility locator so you can find the type of care you need close to your new home.

http://www.floridahealthstat.com
http://www.floridacomparecare.gov

Hospitals

According to U.S. News and World Report's 2006 List of America's Best Hospitals, eight can be found in Florida. These include:

Gainesville
Shands at the University of Florida
http://www.shands.org/hospitals/uf/

Jacksonville
St. Vincent's Medical Center
http://www.jaxhealth.com/
*Note that Jacksonville is also home to a world renowned Mayo clinic
http://www.mayoclinic.org/jacksonville

Miami
Bascom Palmer Eye Institute
http://www.bpei.med.miami.edu/site/default.asp
University of Miami, Jackson Memorial Hospital
http://www.um-jmh.org/

Orlando
Florida Hospital
http://www.floridahospital.org/locations/fhsouth/index.htm

Sarasota
Sarasota Memorial Hospital
  http://www.smh.com/

Tampa
H. Lee Moffitt Cancer Center and Research Institute
  http://www.moffitt.usf.edu/
Tampa General Hospital
  http://www.tgh.org/

VA Medical Facilities

A high percentage of people moving to Florida have served in the military and are entitled to health care benefits from the Veteran's Administration. There are several VA Hospitals and Clinics throughout the state as well as VA Outpatient facilities located near most major cities.

> For a complete list of VA hospitals, clinics, and outpatient facilities visit
> http://www1.va.gov/directory/guide/allstate.asp

ARTS AND CULTURAL ACTIVITIES

If you're into the arts, cultural events, and festivals, Florida will not disappoint you. There are more than 340 museums, more than 30 theatre companies, more than 200 outdoor festivals, and countless galleries and craft shops dedicated to the arts. Most major Florida cities have symphony orchestras. The addition of theme parks, and world-class beaches and state parks means that there is no reason for you or your visiting kids and grandkids to ever be bored. In fact, there is so much to do in all parts of Florida that they should be begging to come visit. And because the weather is so nice, most of these places and events can be visited and enjoyed year round.

MUSEUMS

Let's take a look at some popular museums throughout Florida. Keep in mind this is not even close to being a comprehensive list, just a few highlights.

For a comprehensive list of museums in Florida, visit the Florida Association of Museums website at http://www.flamuseums.org

Ringling Museum of Art

The name should ring a bell if you've ever been to the circus. This museum, containing over 10,000 works of art including paintings, sculpture, drawings, prints, photographs and decorative arts, is located on the west coast of Florida in Sarasota. Created by circus owner John and his wife Mable, the Ringling Museum has been in operation since 1927 and is now run by Florida State University.

For more information visit http://www.ringling.org

Salvador Dali Museum

Visited by over 200,000 people each year, the Salvador Dali Museum is the largest collection of Dali's work in the world. Previously housed in Cleveland, Ohio, the museum opened in St. Petersburg in 1982. The museum has such an extensive collection that it frequently loans its work out to other institutions throughout the world.

For more information visit
http://www.salvadordalimuseum.org

Florida Museum of Natural History

Located in Gainesville at the University of Florida, this is the largest museum of natural history in the southern United States. The Florida Museum of Natural History holds over 20 million specimens of mammals, birds, reptiles, mollusks, fish, butterflies, and fossils. One highlight of the museum is a 6,400 square feet Butterfly Rainforest exhibit.

For more information visit the museum's website at
http://www.flmnh.ufl.edu

Southeast Museum of Photography

Located in Daytona Beach on the campus of Daytona Beach Community College, the Southeast Museum of Photography is one of less than 12 facilities in the United States dedicated exclusively to photography and is the only one in the southern United States. A special cultural and educational resource for anyone interested in photography, the museum features exhibitions covering a wide variety of photographic styles, and lectures featuring prominent photographers and critics.

For more information visit their website at http://www.smponline.org

World Golf Hall of Fame

The website sums it up: "If you love golf, you've got to go!" Previously located in Pinehurst, North Carolina, the World Golf Hall of Fame moved to St. Augustine and a new multi-million dollar facility in 1998. The museum contains historical artifacts and personal memorabilia from all the game's biggest stars. Also on the grounds of the museum is an 18-hole natural grass putting course and an IMAX Theater. There are also two golf courses in what is called World Golf Village. The courses are King and Bear, co-designed by Arnold Palmer and Jack Nicklaus, and Slammer and Squire, designed by course architect Bobby Weed, who consulted with Sam "The Slammer" Snead and Gene "The Squire" Sarazen. The World Golf Village offers "stay and play" packages so that you can visit the museum as well as play the courses while in town.

For more information visit http://www.wgv.com

THEME PARKS

There's perhaps no better way to spend quality time with your kids and grandkids when they come to Florida than to visit one of the many theme parks that Florida has to offer. Most of Florida's theme parks are located in central Florida, making them easy to get to from almost anywhere in the state. There are a few things you need to know, however, before you go.

Price of Admission

Admission prices change frequently, though you are easily looking at over $100 for two people at most of the parks listed below. Discounts are usually available for Florida residents, senior citizens, members of the military, AAA members, and young children. Annual passes are also available and can provide good savings if you plan to visit a park several times a year. Visit the parks' websites for current admission prices and information on any discounts currently available.

When Not To Go

The best times to avoid the theme parks are during the summer, spring break, or winter break when millions of kids are out of school and family vacations are underway. Nothing's worse than waiting in line for hours on end in 95 degree heat, packed in with thousands of people, especially when you are seemingly the only person there who remembered to apply deodorant that morning. Also, try to steer clear of most major holidays. If you plan to live in Florida at least part time or visit frequently, this still leaves you with plenty of time to enjoy the theme parks at times when they are less crowded.

Theme Park Guides

If you would like more information on Florida's theme parks, two of the best guides are Lonely Planet: Orlando & Central Florida by Wendy Taylor (Lonely Planet Productions, 2003) and Frommer's Walt Disney World and Orlando 2007 by Laura Lea Miller (Frommer's 2006). Both are available at amazon.com or ask for them at your local bookseller.

Walt Disney World's Magic Kingdom

The granddaddy of them all, the house of the mouse, is what got this whole Florida tourism based economy started. Walt Disney World's Magic Kingdom, which opened for the first time in 1971, is one of the most visited and famous theme parks in the world. Enjoyed by millions of youngsters, adults, and championship winning sports teams alike ("I'm going to Disney World!"–Various), Walt Disney World's Magic Kingdom should be at the top of your list of things to see and do, at least once, while in Florida.

The park is open 365 days a year and features—in addition to the rides of course—several parades, shows, and exhibits geared towards the enjoyment of the whole family.

Epcot

Mickey's neighbor, Epcot Center is equally enticing with its cultural charm. Epcot features rides like Mission: Space, and exhibits as well as festivals throughout the year such as the International Flower and Garden Festival, and the International Wine and Food Festival. There's something for everyone at Epcot, as the park is divided into different foreign country themed "pavilions," such as China, Mexico, Germany, France, and more. Many people make a special trip to Epcot just to eat, wanting to sample cuisines from all over the world in one location.

MGM Studios

Also part of the Disney family of theme parks, MGM Studios offers its visitors the chance to immerse themselves in their favorite movies. They have rides based on movies such as Star Wars, and Twilight Zone Tower of Terror as well as shows based on hit films such as Beauty and the Beast and the Little Mermaid.

Walt Disney World's Animal Kingdom

Disney's newest addition to its theme park lineup is perhaps its most exciting. Celebrating the wonders of nature and wildlife, Animal Kingdom has rides, shows, and attractions that rival those at its older brother and sister parks. In addition to several animal encounters, rides such as Kilimanjaro Safaris, Kali Rapids, and the new Expedition Everest will make a trip to Animal Kingdom worth your while.

> For more information on all of Walt Disney World's parks, hotels and other attractions visit:
> http://disneyworld.disney.go.com/wdw/index

Universal Studios/Islands of Adventure

Just down the road from Walt Disney World are the formidable competitors Universal Studios and Islands of Adventure. Here you will find two distinct movie themed parks, with a "City Walk" in between them, flanked by three exciting themed hotels—The Portofino Bay, The Hard Rock, and The Royal Pacific—all connected by a unique and ultra-cool ferry system. It all combines to create one of the most fun environs in all of Florida. World famous rides such as the Incredible Hulk and Revenge of the Mummy provide the thrills, while staples such as Spider-Man and Shrek 4-D provide all-ages entertainment. At Universal City Walk there is an Emeril's and Jimmy Buffet's Margaritaville restaurants and other eating and drinking establishment such as Pat O'Brien's and City Jazz. A two-story, 20 screen Loews movie theatre provides yet another opportunity to be entertained. My advice is to reserve a room at one of the hotels mentioned above and take a couple days to experience all Universal Studios, Islands of Adventure, and City Walk have to offer. Plus, with your room key you can gain special front of the line access to the rides in both parks as well as discounts at some of the restaurants and shops.

> For more information on Universal Orlando's theme parks, hotels, and entertainment visit http://www.universalorlando.com

Sea World

Located in Orlando, Sea World is one of the most famous marine life oriented theme parks in the world. There are shows and attractions featuring various forms of sea life such as dolphins, penguins, stingrays, sharks, killer whales, and more. If these shows aren't exciting enough, the park has added rides and roller coasters in recent years for guests of all ages to enjoy.

> For more information regarding Sea World,
> its hours of operation and other information visit
> http://www.seaworld.com/seaworld/fla/default.aspx

Marineland: The Original Sea Park

Marineland first opened in the late 1930's as the world's first oceanarium, providing visitors the first glimpse available of ocean life here on dry land. It was originally called "Marine Studios" and served as the site for filming of several Hollywood productions. Today, Marineland is a research and education facility, but the park itself is still open to the public. The main attraction at Marineland is the dolphin show. Marineland is located on the east coast of Florida, just south of St. Augustine. It might not warrant a special trip from far away, but if you are in the area, it's worth your time to see it.

> For more information visit http://www.marineland.net

Busch Gardens Florida

If you are looking for the adventure of an African Safari but don't want the hassle of updating your passport, Busch Gardens Tampa Bay may be a great alternative. Like most other theme parks, Busch

Gardens also has rides, roller coasters, shows, and attractions but all with a Safari-themed twist. Most visitors though, go for the animals, and there are plenty there for you to see.

> For more information on Busch Gardens visit
> http://www.buschgardens.com/BGT/default.aspx

Kennedy Space Center

Located on Florida's East Coast near Cocoa, Kennedy Space Center is the site of all United States space shuttle launches as well as the launch of many military and civilian rockets carrying satellites. If you've never seen a shuttle launch in person, put it on your list of things to do, and do it. There's nothing quite like it. There is also a visitors' complex featuring exhibits, shows, and other attractions.

> For details about Kennedy Space Center,
> as well as launch schedules visit
> http://www.kennedyspacecenter.com/

FESTIVALS

Most every weekend in Florida offers a wide range of events and activities in which you may partake. If you're not too busy visiting the beaches or enjoying the parks and other attractions, be sure to check out some of these exciting festivals throughout the state.

Strawberry Festival

Every year Floridians in-the-know flock to Plant City, located between Tampa and Orlando, to partake in some good old-fashioned fun. The Strawberry Festival is much like a county fair in that it offers amusement rides and games, shows, big-name entertainment, and of course, all the strawberry shortcake you could ever hope to eat. Fortunately, you are not just limited to shortcake, as you will find strawberry ice cream, strawberry sundaes, strawberry milkshakes, strawberry cobbler, and the list goes on and on. It's one festival you've got to check it out, at least once.

> For more information on the Florida Strawberry Festival
> visit the festival website at
> http://www.flstrawberryfestival.com

Florida International Festival

Once every two years Daytona Beach plays host to the Florida International Festival, a musical event featuring symphonic and concert artists from all over the world. The Festival's centerpiece, concerts performed by the London Symphony Orchestra, draws attendees from all over the country to see music at its best.

For more information visit http://www.fif-lso.org

Winter Park Sidewalk Art Festival

One of the nation's most prestigious outdoor art festivals takes place each year in the city of Winter Park, on the outskirts of Orlando. More than 350,000 visitors flock to the show to see almost 1,500 artists showcase their work and compete for cash prizes in a variety of categories. In addition to art, the show also features music and entertainment, food, and children's activities.

Visit http://www.wpsaf.org for more information.

Florida Film Festival

It's not quite Sundance, but some would say that's a good thing. Every year in Orlando, filmmakers and film lovers alike come together to celebrate the world of film in what has become "one of the most respected regional film events in the country." In addition to screenings of movies, the festival includes seminars, educational forums, glam parties, and other special events.

For more information, visit the festival website at http://www.floridafilmfestival.com

PARKS

The Florida Park Service, managed under the Florida Department of Environmental Protection, runs one of the largest park systems in the country with 159 parks spanning more than 723,000 acres and 100 miles of beaches. Activities available for you to enjoy include swimming, diving, or snorkeling in Florida's rivers and springs, bird watching, fishing, and hiking on scenic nature trails. Florida's parks, combined with wonderful weather, offer year-round fun for all ages. Events such as battle reenactments and Native American festivals celebrate Florida's past, while art shows, museums and lighthouses offer a look into Florida's cultural heritage.

> For more information on Florida's park system and a comprehensive list or parks visit http://www.floridastateparks.com

Dry Tortugas National Park

It's not the easiest place to get to, but if you're up for an adventure, take a day trip to Dry Tortugas National Park. Considered part of the Florida Keys, the Dry Tortugas are actually closer to Cuba than they are the United States. Discovered in 1513 by Juan Ponce de Leon, the park can only be reached by seaplane or boat.

There is plenty to see and do at the Dry Tortugas National Park, including snorkeling, fishing, boating, and touring the site's historic Fort Jefferson.

> Visit the park's website for more information at http://www.nps.gov/drto

BEACHES

When one's thoughts turn to Florida, the first image that comes to mind for many people is a beach: white sand, gentle breeze, and the hypnotic sounds of the lapping of the waves. Ah, this is why we live here. Florida has more than 1,100 miles of coastline, the majority of that being white sandy beaches bathed in glorious sunshine for you to enjoy. Each beach in Florida is unique, and you are sure to enjoy visiting several different beaches around the state to see which is your favorite. Whether your pastime is surfing, boating, kayaking, or just floating around, you're sure to find a beach that fits your mood close by.

> To learn more about Florida's beaches and view an interactive guide visit: http://www.visitflorida.com/experience/beaches/

LIGHTHOUSES

People are continually enchanted by the history, lore, and romance embodied by lighthouses. Visiting and learning about lighthouses has become a passion for people of all ages. Being that it's nearly surrounded by water, Florida has a large number of lighthouses. Many are open to the public for tours, and some are even available to climb.

According to the Florida Lighthouse Association, there are 30 remaining historic lighthouses in Florida. Some of these are among the nation's oldest and tallest, such as Ponce de Leon Inlet Lighthouse (2nd tallest in U.S.), the only Florida lighthouse registered as a National landmark. The Florida Lighthouse Association, whose mission is to preserve Florida's remaining lights, offers some great information on the history of lighthouses in Florida at the website http://www.floridalighthouses.org. If you are interested in seeing these maritime marvels first-hand, then surf on over to http://www.visitflorida.com/cms/d/floridas_lighthouse_trail.php.

FISHING AND BOATING

Not many places in the world, let alone in the United States, can beat Florida when it comes to the quality of fishing and boating. In fact, with 7,700 lakes, 10,550 miles of rivers, and 2,276 miles of tidal shoreline, Florida is the "fishing capital of the world" and some would consider it the boating capital of the world also. Florida has a large variety of species of fish, from largemouth bass in the fresh waters, to redfish along the shoreline to sailfish offshore. More anglers come to Florida to fish than anywhere else in the nation.

Florida has the third highest number of boat owners in the nation, ranking behind Michigan and California. And why wouldn't it, with water everywhere you turn? Boating is a favorite recreational pastime of many Floridians and visitors to the state and an excellent way to relax and spend time with friends and family. No matter where you are in Florida: North, South, East or West, inland or on the coast, good fishing and boating are just outside your door.

> For more information on fishing and boating in Florida, as well as license information for both, visit: http://www.fishingcapital.com and http://www.myfwc.com/boating.

PROFESSIONAL SPORTS

Most everyone enjoys a live sporting event now and then, and it will be an even more special experience for your kids and grandkids when they can experience it as part of their visit to you in Florida. Some of Florida's professional teams include:

NFL

Jacksonville Jaguars
Miami Dolphins
Tampa Bay Buccaneers

NBA

Miami Heat
Orlando Magic

MLB

Florida Marlins
Tampa Bay Devil Rays
Major League Baseball Spring Training is also a big hit in Florida, with 18 teams calling the sunshine state home in February and March.

NHL

Florida Panthers
Tampa Bay Lighting
Florida is also home to the Professional Golf Association (PGA) of America headquarters (Palm Beach Gardens), the Ladies Professional Golf Association (LPGA) headquarters (Daytona Beach), the Association of Tennis Professionals (ATP Tour) headquarters (Ponte Vedra), and the National Association for Stock Car Auto Racing (NASCAR) corporate headquarters (Daytona Beach).

> For more information on professional sports in Florida and links to all teams, visit the Florida Sports Foundation's website at http://www.flasports.com. Once there, you can also request free guides published by the Foundation for information on golf, fishing and boating, spring training and more.

GLOSSARY

No book about real estate would be complete without the obligatory real estate glossary. No matter how many homes you've bought or sold, there always seem to be new words showing up in the industry that you may not have come across yet. Here is a selection of the most often used terms and jargon in the industry today.

Abstract of Title A version of the title history of a given property, be it a house, condo, or just a piece of land. It lists any changes of ownership (sales), and any mortgages or liens that may have been or are currently connected with the property.

Acceleration Clause If you as the borrower default on your mortgage by failing to make your payments, the lender has the right to demand payment of the entire outstanding loan balance by invoking this clause.

Acre A measurement of land that equals 4,840 square yards or 43,560 square feet.

Adjustable Rate Mortgage (ARM) These are mortgage loans that start off with a rate typically lower than that of a fixed-rate mortgage, but adjust after a certain period of time. Whether they adjust up or down depends on what interest rates are doing at the time. The rate is usually pegged to a certain index such as t-bills and there are usually limits as to how much the rate can fluctuate. If you see an advertisement for a 5/1 ARM, this means that the initial interest rate is fixed for the first 5 years and then adjusts accordingly for the remainder of the life of the loan.

Ad Valorem Simply means "according to value." You will usually see this term in the discussion of taxes and tax assessments.

Amortization A typical mortgage payment is made up of principal and interest. Over the life of the loan, the interest part decreases and the principal increases so that the loan amount is paid off in the required time.

Annual Percentage Rate (APR) This is an interest rate that reflects what the mortgage will cost as a yearly rate. When shopping for a mortgage, you will see a rate, followed by an APR. The APR will usually be higher, as the APR includes points as well as other front end loan costs. When shopping for a mortgage, don't just look at the rate. Pay close attention to the APR.

Appraised Value The appraised value is an appraiser's estimate of the value of the property.

Appreciation An increase in the value of a property. Appreciation can be due to changes in the market or improvements to the property.

Assessment An assessment is a tax levied on a property to pay for things like road improvements, sewers, and streetlights.

Assignment When a real estate contract or a mortgage is transferred from one person to another, it is called an assignment.

Assumable Mortgage This is a mortgage that can be taken over by the buyer of the home. Assumable mortgages, while not as popular today as they once were, can be beneficial to both the buyer and the seller in the right situation.

Back-End Ratio Used in the mortgage underwriting process, this formula is simply your monthly debt divided by your gross income.

Balloon Mortgage A short-term loan with low monthly payments for a certain period of time and one large payment at the end.

Bi-Weekly Mortgage A mortgage that requires you to make payments every two weeks instead of once a month. With a bi-weekly mortgage, instead of making 12 full payments a year, you end up making 13, reducing the principal amount, as well as the time it takes for repayment.

Blanket Mortgage This is a mortgage that covers more than one piece of real estate.

Blueprints The set of plans consisting of drawings, diagrams, measurements and dimensions that the builder and each subcontractor will use to build your home.

Breach of Contract A break or violation of the terms of a real estate agreement.

Bridge Loan A means of financing the purchase of a property without having to sell one's current home. Because rates are higher than normal and they can require the buyer to make two or more mortgage payments, bridge loans should only be used in special circumstances.

Buy Down A buy down is when the borrower makes a lump sum payment up front in order to "buy down" the rate so that they can have lower payments for an initial period. Most of the time this lump sum will come from the seller as a means of getting someone to buy his or her property instead of another property.

Buyer's Agent A real estate agent who only works on behalf of the buyer in a given real estate transaction.

Cash Flow The amount of cash generated over certain period of time from an income producing property such as a rental unit. Positive cash flow occurs when the amount of income generated exceeds the amount of the expenses. Negative cash flow, on the other hand, is when the expenses exceed the cash flow.

Certificate of Eligibility Document that is given to qualified armed services veterans that entitles them to VA loans.

Certificate of Reasonable Value (CRV) The VA version of an appraisal.

Certificate of Title Confirms that the current owner legally holds title to a property.

Chain of Title The history of all the title transfers of a piece of property.

Chattel A legal term for personal property.

Closing The final exchanging of the property and money between the buyer, seller, and if applicable, the lender.

Closing Costs These are all the costs incurred by the buyer and seller in a real estate transaction. Examples are taxes, title insurance, credit report fees and recording fees.

Closing Statement A summary of the financial details of a real estate transaction. The closing statement will show how much money is due to or from the seller and buyer, any mortgages, and all closing costs related to the transaction.

Cloud A cloud is something found by the title search which shows that the property is not owned free and clear by the seller.

Commitment A promise from a lender to lend money.

Comparables "Comps" Used in a comparative market analysis or appraisal, these are properties in close proximity and of similar size and amenities as the subject property.

Comparative Market Analysis Used primarily as a tool by real estate agents to help sellers price their homes correctly, it can also be used by buyers to find out how much they should pay for a home by comparing the selling prices of comparable homes in the immediate area.

Condominium Buildings in which the owners own separate units of the building and share responsibility for, and use of, the common areas.

Construction Loan This is a loan to pay for the construction of a home, condo, or other building. Instead of all the money being loaned up front, the lender will disburse the funds at specific intervals during the construction process.

Contingency A condition that must be met before a contract becomes completely binding. Examples of common contingencies include getting a loan, home inspections, or home to sell contingencies.

Contract of Sale The agreement between the buyer and seller that outlines the conditions of the transfer of title to the property, such as sales price and purchase terms.

Conventional Loan This is a mortgage that is obtained in the open market, not guaranteed by the VA, nor insured by the FHA.

Conveyance A document that transfers ownership of a property from one person to another. Examples of conveyances are deeds and leases.

Covenants, Conditions and Restrictions (C, C & R's) Set of documents that includes the rules that govern a Homeowners' or Condo Association.

Cul-de-Sac These are dead end streets with areas to turn around at the end, common in many subdivisions. Cul-de-sacs have been popular in the past because they were thought to cut down on through traffic and speeding but recent research suggests they adversely affect the traffic patterns in the rest of the community. As a result, they are being used less and less in newer subdivisions.

Debt-to-Income Ratio A borrower's monthly payments on long-term debts divided by their gross monthly income.

Deed This is the actual document that transfers the title to real property. The two main types are quitclaim deed and warranty deed.

Deferred Interest This is unpaid interest added to the balance of a loan.

Documentary "Doc" Stamps Usually paid for by the seller, doc stamps cost 70 cents per $100 of the sales price.

Due-on-Sale Clause This clause allows the lender to call the loan due to be paid should the property change ownership (be sold).

E-Pro Realtors who have completed specialized training in using technology to the betterment of real estate transactions.

Earnest Money The deposit given by a buyer to the seller in order to form a binding contract.

Easement This is a right of way that gives access to people other than the owner of the property. Common examples of easements are utility easements and drainage easements.

Encroachment An encroachment is an illegal intrusion on someone else's property. For example, if you build a fence beyond your property line onto your neighbor's property, that is an encroachment.

Glossary

Encumbrance This is simply a lien or a claim on a property.

Equal Credit Opportunity Act (ECOA) Federal law that makes it illegal for lenders to discriminate against borrowers based on the borrower's race, color, religion, age, sex, national origin, marital status, or their dependence on public assistance.

Equity The value of a property minus any debts owed on the property.

Escrow These are funds that are set aside, usually by your mortgage lender, to pay for expenses such as taxes and insurance.

Federal Home Loan Mortgage Corporation, "Freddie Mac" (FHLMC) Agency that buys conventional mortgage loans from savings and loans and approved mortgage bankers.

Federal Housing Administration (FHA) This agency insures residential mortgages made by private lenders. The FHA is a division of the Department of Housing and Urban Development (HUD).

Federal National Mortgage Association, "Fannie Mae" (FNMA) Freddie's sister, Fannie Mae is a corporation started by Congress that buys and sells different types of mortgages, including conventional, FHA, and VA loans.

FICO Score This is a score ranging from 300 to 850 that lenders will use to determine your credit risk, or your ability to repay a loan.

Fixed-Rate Mortgage With a fixed-rate mortgage the interest rate is set or "locked-in" for the life of the loan.

Flood Insurance A flood insurance policy is a type of insurance policy that everyone should consider, and is required by lenders if your property is in a flood zone determined to be at a high risk for flooding.

Foreclosure If you fail to make payments on your mortgage, the lender can force the sale of your property through the legal process of foreclosure.

Front-End Ratio This formula (calculated by taking monthly mortgage payments divided by your gross monthly income) determines how much of your income goes to repaying a loan.

GRI (Graduate of the Realtor Institute) REALTORS® who have earned this designation have completed 90 hours of specialized training and testing and can be considered among the most well-trained real estate practitioners.

Government National Mortgage Association, "Ginnie Mae" (GNMA) This is a government agency that provides funds for FHA and VA mortgages.

Guaranteed Mortgage A mortgage that is guaranteed by a third party, such as a VA mortgage.

Hazard Insurance This is a type of insurance that insures against certain types of loss such as fire, flood, lightning, or wind damage. Together with personal liability insurance and any other mandatory (flood in some cases) or optional coverage, this will form your homeowner's insurance policy.

Home Equity Line of Credit (HELOC) This is a loan against the equity you have in a property.

Home Inspection A home inspection is recommended on all home purchases, but especially on homes being resold by someone other than a builder providing a full warranty. A home inspection should cover all major systems (electrical, plumbing, heating and air-conditioning), the physical condition of the property, as well as all structural elements.

Homeowner's Insurance Insurance that is required by most lenders that combines personal liability insurance with hazard insurance, as well as any other mandatory or optional coverage, to provide protection of a property as well as its contents.

Homeowner's Warranty A warranty that covers repairs or replacement of specific parts or systems of a house for a specific amount of time. Most new homes come with a warranty provided and paid for by the builder. Sellers of resale homes will also often times provide a warranty through a third-party warranty company as an incentive for someone to purchase their home.

HUD-1 This is a standardized closing statement form listing all costs, credits, fees, escrow amounts, etc. involved in a real estate transaction.

HVAC Stands for Heating, Ventilation and Air Conditioning.

Impact Fees Fees levied by a city or municipality to help pay for things like new roads, new schools and other infrastructure required to support the growth of an area. Usually these fees are paid for directly by the builder/developer, but they are passed on to homebuyers in the total price of the house.

Interest The amount of money charged on the amount of money borrowed.

Joint Tenancy Popular way for unmarried people to take title to a property. When one owner dies, the surviving owner becomes the sole owner of the property.

Jumbo Loan A jumbo loan is a loan that is higher than the limits set by Fannie Mae and Freddie Mac for conventional loans. For 2006 the limit is $417,000.

Lease-Option A lease-option is a great way for boomers, or anyone for that matter, to try out a certain home or certain area before committing completely to buying. In a lease-option, the buyer agrees to initially rent the property for a certain period of time and at the end of the lease period they have the option to purchase the property, usually at a price determined when the lease-option started.

Lease-Purchase Not to be confused with a lease-option, a lease-purchase obligates you to purchasing the property, whereas with the lease-option you have a choice. In a lease-purchase, a certain amount of the rent being paid can go to paying either principal, taxes, insurance, or the downpayment, or a combination of all those.

Glossary

Loan-to-Value Ratio (LTV) This is the amount of the mortgage loan, divided by the value of the property.

Margin Used with adjustable rate mortgages, the lender will add the margin to the index to determine your adjusted interest rate.

Market Value The expected amount that a property would sell for in an open market.

Mortgage Most easily explained as a lien against a property until it is paid for.

Mortgage Banker A mortgage banker is someone who originates or provides mortgages to later resell in the secondary market.

Mortgage Broker A mortgage broker helps to match individuals looking to borrow money with loans from many different possible sources. They are usually paid a commission for their services, but sometimes receive just a flat fee.

Mortgage Insurance The amount of money you pay to insure the mortgage when your down payment is less than 20 percent.

Mortgage Insurance Premium (MIP) Equals 0.5% of the mortgage amount that borrowers of an FHA insured mortgage pay each month as insurance against default.

Mortgagee The party that lends the money. They hold the mortgage.

Mortgagor The party that borrows the money. They give a mortgage to the lender.

Negative Amortization Sometimes referred to as "Negative Am", this is when your monthly payment amount is not enough to cover all the interest on the loan. The unpaid interest is added to the principal, resulting in the borrower owing more than the original amount borrowed.

No-Doc Loan No-doc or low-doc loans are types of loans that require very little documentation to get a loan. They usually require large down payments, and are most popular with those who are self-employed or people who for privacy or other reasons do not want to disclose all of their income or assets.

Note A signed obligation to repay a debt.

Origination Fee This is the fee that the lender charges to perform certain tasks such as preparing the loan documents, making your credit check, and arranging for the property to be inspected and/or appraised. The origination fee is usually a percentage of the loan.

Permanent Loan Can be defined as any mortgage loan for a period of 10 years or more.

PITI Acronym for principal, interest, taxes, and insurance that together makes up your total mortgage payment.

Points Points are essentially prepaid interest on a loan. One point equals one percent of the loan amount.

Power of Attorney A power of attorney or "POA," is a legal document that authorizes one person to act on behalf of another.

Prepaid Expenses This is money that is needed to create an escrow account when closing a real estate transaction. Prepaid expenses will usually include taxes, insurance, homeowners' association dues, and special assessments.

Prepayment Penalty This is a fee for the early repayment of a loan. You should do your best to avoid mortgages with this clause in them, unless you are positive that you will not be paying off the loan early.

Principal The principal is the amount of debt remaining to be paid on a loan.

Private Mortgage Insurance (PMI) This is mortgage insurance for conventional loans that is usually required when the amount financed has a greater than 80% loan to value ratio (LTV).

Punch-out A step in the construction process when any remaining items that need to be corrected are taken care of or "knocked-out." Also called "bump-out" or just "bump," there will be several times during the construction of a home that punch-out occurs, for example, masonry bump, framing bump, electrical bump and so on.

Qualification Rate This is the interest rate used to determine whether or not you will qualify for a particular loan. It is not necessarily the rate that you will end up paying.

Quit Claim Deed This is a document that transfers title from one person to another. The person transferring the title is giving up or "quitting" all claims to the property.

R-Value This is the standard of measurement of an insulation's ability to keep heat from entering a home. The R-Value of all insulation installed in new construction homes in Florida is required to be disclosed to you at the time of contract.

Radon A naturally occurring radioactive gas present everywhere but found in high concentrations in some parts of Florida. Radon testing is available and if the seller has any knowledge of previous radon inspection results, they are required to disclose those results to the buyer.

Rate Lock When a mortgage interest rate is kept at a set rate for a specific period of time. Rate locks vary in length. The longer the rate lock, the more it will cost.

REALTOR® A real estate agent or broker who is a member of the National Association of REALTORS® (NAR).

Recision This is the process of canceling a contract. Some contracts, primarily for condominiums, have a recision period, or time frame within which you can cancel the contract.

Glossary

Recording Fees Money in the form of recording fees is required to make the sale of a property public record.

Refinance Used primarily to obtain a better interest rate and therefore lower your payments, or in many cases to pull cash out of a property, refinancing is when you get a new mortgage or replace an existing mortgage on a property you already own.

Real Estate Settlement Procedures Act (RESPA) Federal law that dictates that consumers be allowed to review any known or estimated settlement charges that they will incur at least once after they make application and once before or at the closing.

Reverse Annuity Mortgage (RAM) Essentially the reverse of a conventional mortgage, in this case the lender makes payments to the borrower using the equity in the home as collateral. Though neither very often used nor highly recommended, this can sometimes be a smart financial or estate-planning tool.

Right of First Refusal An agreement that gives one party the chance to buy or lease a property before anyone else has the opportunity.

Servicing The activities a lender must perform to keep a loan in good standing. Activities include sending statements, collecting and processing payments, and paying items like taxes and insurance.

SEER Rating Stands for "Seasonal Energy Efficiency Rating;" this is a measure of how energy efficient an air conditioning unit is. The federal standard is now 13 SEER.

Simple Interest Interest calculated only on the amount of the principal.

Survey A survey, conducted by a registered land surveyor that provides detailed measurements of a property.

Tenancy in Common Popular way to take title for people who are on their second (or more) marriage. When one party dies, their share of the property can be willed to their children or other heirs.

Term The amount of time over which a loan is set to be repaid.

Title The title is a document that declares and proves one's ownership of a particular property.

Title Insurance A title insurance company will issue title insurance to insure a homebuyer against errors in the title search or clouds or encumbrances on the title.

Title Search A search of public records by a title company or an attorney to determine who has legal ownership of a property.

Truth-in-Lending Law Requires that the APR of a loan be disclosed to borrowers shortly after they apply for a mortgage.

Underwriting This is the process conducted by a lender to determine whether or not to loan money to a particular borrower, and if so, at what rate and terms. Takes into account a buyer's credit history, assets, length and type of employment, and other factors.

VA Loan A loan guaranteed by the Department of Veterans Affairs, available to veterans whose length and type of service qualify them for this benefit.

Vantage Score A new credit scoring method that combines the scores from Equifax, Experian, and Transunion to create an easier method for consumers, as well as lenders, to determine what their credit score actually is.

Walk-Through The last step before closing in the new construction process. You will walk through your home with the builder or a customer service representative, going over how everything functions and looking for anything that needs to be corrected or fixed. You should also do a walkthrough before any other real estate closing, especially if there has been any significant time lapse between your home inspection and the closing, in order to make sure everything you contracted for is in place and also to make sure the seller or previous tenant didn't do any damage when they moved out.

Warranty Deed This is a type of deed where the seller guarantees that he holds clear title to a property and therefore has a right to sell it to a buyer.

Zoning Laws City restrictions regarding what kind of properties can be built in particular locations. A city will be divided into different zones such as commercial and residential, and even more specialized zones of each such as light commercial and multi-family. Zoning laws also dictate what size the structures must be as well as where they should be located within the particular zone.

WEBSITE INDEX

You will find direct links to all of the sites listed here on the "Florida Resources" page at FloridaforBoomers.com

CHAPTER 1 AN INTRODUCTION TO FLORIDA

My Florida
http://www.myflorida.com

CHAPTER 2 CHOOSING AN AREA

Florida Resources - Florida for Boomers
http://www.floridaforboomers.com/resources (Password-protected)
(The password is on pages xvi and 7)

Google Groups
http://groups.google.com

Yahoo Groups
http://groups.yahoo.com

CNN/Money Magazine's Best Place Tool
http://money.cnn.com/magazines/moneymag/bpretire/2006/FL.html

Realtor.com
http://www.realtor.com

Zillow.com
http://www.zillow.com

Homes and Land Magazine
http://www.homesandland.com

The Real Estate Book
http://www.therealestatebook.com

Digest of Homes
http://www.digestofhomes.com

Where to Retire Magazine
http://www.wheretoretire.com

Living Southern Style
http://www.livesouth.com

Second Home Journal
http://www.2ndhome.net

Live South Real Estate Shows
http://www.livesouthshows.com

Florida Lifestyle Shows
http://www.floridaliving.org

Find an Agent – Florida for Boomers
http://www.floridaforboomers.com/agents

CHAPTER 3   HOW TO FIND A REAL ESTATE AGENT

DBPR Online Services (Search for licensee complaints)
https://myfloridalicense.com

CHAPTER 4   TYPES OF HOMES IN FLORIDA

Mark Zilbert – South Florida Condos
http://www.zilbert.com

Florida Manufactured Housing Association
http://www.fmha.org

CHAPTER 5   TYPES OF COMMUNITIES IN FLORIDA

Florida's Official Golf Website
http://www.playfla.com

Florida Golfer Guide
http://www.floridagolferguide.com

List of Registered 55-Plus Communities in Florida
http://fchr.state.fl.us/55+_registered_list.htm

Del Webb Communities – Wiregrass Country Club
http://www.delwebb.com

Website Index

The Palms – Maintenance-Free Community
http://www.winston-james.com

Villages of Royal Palm – Maintenance-Free Community
http://www.villagesofroyalpalm.com

The Ginn Company – Reunion Resort and Club
http://www.reunionresort.com
http://www.ginncompany.com

CHAPTER 6 HOMEOWNER'S AND CONDOMINIUM ASSOCIATIONS

Florida Condo and Homeowners' Association Law – Glazer and Associates, P.A.
http://www.condo-laws.com

Community Development Districts
http://www.floridaspecialdistricts.org

CHAPTER 7 PROPERTY TAXES

Property Appraiser's Offices throughout Florida
http://myflorida.com/dor/property/appraisers.htm

Florida Property Tax Reform Committee
http://www.propertytaxreform.state.fl.us

CHAPTER 8 HOMEOWNER'S INSURANCE

Florida Property and Casualty Insurance Reform Committee
http://www.myfloridainsurancereform.com

Florida Market Assistance Program
http://www.fmap.org

Citizen's Property Insurance Corporation
http://www.citizensfla.com

My Safe Florida Home - Insurance Discounts
http://www.mysafefloridahome.com/insurance.asp

National Flood Insurance Program
http://www.floodsmart.gov

CHAPTER 9 CONTRACTS AND DISCLOSURES

Coastal Construction Control Line
http://www.floridadep.org/beaches

Florida Building Energy Efficiency Rating System
http://www.dca.state.fl.us/fbc/committees/energy/EnergyBrochure-110602.pdf

Florida Homeowner's Construction Recovery Fund – Construction Industry Licensing Board
http://www.myflorida.com/dbpr/pro/cilb/cilb_index.shtml

CHAPTER 10 HOME INSPECTIONS AND WARRANTIES

Amerispec – Home Inspections
http://www.amerispec.com

Pillar to Post – Home Inspections
http://www.pillartopost.com

American Home Shield – Home Warranties
http://www.ahswarranty.com

Old Republic Home Protection
http://www.orhp.com

CHAPTER 12 REAL ESTATE FINANCING

1031 Exchange
http://www.starker.com

FICO Score
http://www.myfico.com

Vantage Score
http://www.vantagescore.com

Annual Credit Report
http://www.annualcreditreport.com

Mortgage Calculators – Florida for Boomers
http://www.floridaforboomers.com/mortgagecalculators

CHAPTER 13 NEW HOME CONSTRUCTION

Security System Monitoring Companies

ADT
http://www.adt.com

Brinks
http://www.brinks.com

Alarm.com
http://www.alarm.com

CHAPTER 18 FLORIDA RESOURCES

Hurricanes

National Hurricane Center – Saffir-Sampson Scale
http://www.nhc.noaa.gov/aboutsshs.shtml

Red Cross
http://www.redcross.org

Federal Alliance of Safe Homes
http://flash.org

My Safe Florida Home
http://www.mysafefloridahome.com

Getting Around

Florida Department of Transportation
http://www.dot.state.fl.us

511 Traffic Info
http://www.FL511.com

Florida Department of Highway Safety and Motor Vehicles
http://www.hsmv.state.fl.us

Driving and the Elderly
http://www.floridagranddriver.com

Distances Between Major Cities
http://www3.dot.state.fl.us/mileage/

Employment

Monster.com
http://www.monster.com

HotJobs.com
http://www.hotjobs.com

CareerBuilder.com
http://www.careerbuilder.com

AgingHipsters.com
http://www.aginghipsters.com. Click on "Careers"

Community Service

GetInvolved.org
http://www.getinvolved.org

Governor's Commission on Volunteerism and Community Service
http://www.volunteerflorida.org

Continuing Education

Florida Colleges and Universities
http://www.fldcu.org

Florida Department of Education
http://www.fldoe.org

Health Care

MyFloridaRx.com
http://www.myfloridarx.com

FloridaHealthStat.com
http://www.floridahealthstat.com

FloridaCompareCare.gov
http://www.floridacomparecare.gov

Shands at the University of Florida
http://www.shands.org/hospitals/uf/

St. Vincent's Medical Center
http://www.jaxhealth.com/

Mayo Clinic - Jacksonville
http://www.mayoclinic.org/jacksonville/

Bascom Palmer Eye Institute
http://www.bpei.med.miami.edu/site/default.asp

University of Miami, Jackson Memorial Hospital
http://www.um-jmh.org/

Florida Hospital
http://www.floridahospital.org/locations/fhsouth/index.htm

Sarasota Memorial Hospital
http://www.smh.com/

H. Lee Moffitt Cancer Center and Research Institute
http://www.moffitt.usf.edu/

Tampa General Hospital
http://www.tgh.org/

VA Medical Facilities
http://www1.va.gov/directory/guide/allstate.asp

Museums

Comprehensive List of Florida Museums
http://www.flamuseums.org

Ringling Museum of Art
http://www.ringling.org

Salvador Dali Museum
http://www.salvadordalimuseum.org

Florida Museum of Natural History
http://www.flmnh.ufl.edu

Southeast Museum of Photography
http://www.smponline.org

World Golf Hall of Fame
http://www.wgv.com

Theme Parks

Walt Disney World's Park, Hotels, and Attractions
http://Disneyworld.disney.go.com/wdw/index

Universal Studios/Islands of Adventure
http://www.universalorlando.com

Sea World
http://www.seaworld.com/seaworld/fla/default.aspx

Marineland
http://www.marineland.net

Busch Gardens
http://www.buschgardens.com/BGT/default.aspx

Kennedy Space Center
http://www.kennedyspacecenter.com

Festivals

Strawberry Festival
http://www.flstrawberryfestival.com

Florida International Festival
http://www.fif-lso.org

Winter Park Sidewalk Art Festival
http://www.wpsaf.org

Florida Film Festival
http://www.floridafilmfestival.com

Parks, Beaches and Lighthouses

Florida's Park System
http://www.floridastateparks.com

Dry Tortugas National Park
http://www.nps.gov/drto

Florida's Beaches
http://www.visitflorida.com/experience/beaches/

Florida Lighthouse Association
http://www.floridalighthouses.org

Florida Lighthouse Trail
http://www.visitflorida.com/cms/d/floridas_lighthouse_trail.php

Fishing, Boating and Sports

Fishing
http://www.fishingcapital.com

Boating
http://www.myfwc.com/boating

Florida Sports Foundation
http://www.flasports.com

RECOMMENDED READING

The following books are available at most major booksellers, both online and off.

FLORIDA

Weird Florida by Charlie Carlson (Sterling, 2005)

Frommer's Florida 2007 by Lesley Abravanel and Laura lea Miller (Frommer's, 2006)

Where to Retire in Florida by Richard and Betty Fox (Vacation Publication, 1999)

Choose Florida for Retirement by James F. Gollattscheck and Daniel Murray (Globe Pequot, 2004)

REAL ESTATE

Home Buying for Dummies by Eric Tyson and Ray Brown (Wiley, 2006)

Mortgages for Dummies by Eric Tyson and Ray Brown (For Dummies, 2004)

Home Buyer's Checklist by Robert Irwin (McGraw-Hill, 2001)

RETIREMENT LIVING

The New Retirement: The Ultimate Guide to the Rest of Your Life by Jan Cullinane and Cathy Fitzgerald (Rodale, 2004)

WEALTH BUILDING

Seven Years to Seven Figures: The Fast Track Plan to Becoming a Millionaire by Michael Masterson (Wiley, 2006)

The Automatic Millionaire by David Bach (Broadway, 2003)

Getting Rich Your Own Way: Achieve All Your Financial Goals Faster Than You Ever Thought Possible by Brian Tracy (Wiley, 2004)

BIBLIOGRAPHY

CHAPTER 1 AN INTRODUCTION TO FLORIDA

Wikipedia.org "Florida" < http://en.wikipedia.org/wiki/Florida>
 (accessed September 17, 2006)
Florida Department of Revenue "Intangible Personal Property Tax"
 <http://dor.myflorida.com/dor/taxes/ippt.html>
 (accessed January 10, 2007)
South Florida Water Management District "District Rainfall Data"
 <http://www.sfwmd.gov/site/index.php?id=48>
 (accessed October 21, 2006)
South Florida Water Management District "Current Conditions and
 Forecast" <http://www.sfwmd.gov/site/index.php?id=46>
 (accessed October 21, 2006)

CHAPTER 3 HOW TO FIND A REAL ESTATE AGENT

National Association of REALTORS "When is a Real Estate Agent a
 REALTOR®?" <http://www.realtor.org/realtororg.nsf/pages/
 whoisarealtor> (accessed February 5, 2006)
Florida Association of REALTORS *No Brokerage Relationship Disclosure*
 (Rev. 7/06)
Florida Association of REALTORS *Single Agent Notice* (Rev. 7/06)
Florida Association of REALTORS *Transaction Broker Notice* (Rev. 7/06)
National Association of REALTORS "Why Choose a REALTOR® With a
 GRI designation?" <http://www.realtor.org/griclear.nsf/pages/whygri?>
 (accessed February 5, 2006)

National Association of REALTORS "Why Become a GRI?" <http://www.realtor.org/griclear.nsf/Pages/BecomeGRI?> (accessed February 5, 2006)
The Council of Residential Specialists "Profile of a CRS Designee" <http://www.crs.com/Print/57?> (accessed February 5, 2006)
The Council of Residential Specialists "Why use a CRS" <http://www.crs.com/Print/56?> (accessed February 5, 2006)
The Real Estate Buyer's Agent Council "Frequently Asked Questions about the ABR Designation" <http://www.rebac.net/Content.aspx?PageName=ABRfaq.htm#A1> (accessed January 7, 2007)
National Association of REALTORS "The Accredited Buyer Representative Designation" <http://www.realtor.org/realtororg.nsf/pages/abrdesignation> (accessed September 9, 2006)
National Association of REALTORS "The NAR Resort and Second Home Specialty FAQ" <http://www.realtor.org/resortsweb.nsf/pages/resortq&a> (accessed February 5, 2006)
e-PRO "What is an e-PRO – and why should you use one?" <http://www.epronar.com/whyusepro.htm> (accessed February 5, 2006)

CHAPTER 4 TYPES OF HOMES IN FLORIDA

Palm Harbor Homes "Manufactured or Modular Construction?" <http://www.palharbor.com/our-construction/mfg-or-mod-construction/> (accessed September 30, 2006)
Palm Harbor Homes "Frequently Asked Questions" <http://www.palmharbormarketing.com/faqs.html> (accessed September 30, 2006)

CHAPTER 5 TYPES OF COMMUNITIES IN FLORIDA

Florida Commission on Human Relations "Housing for Older Persons" <http://fchr.state.fl.us/housing_law.htm> (accessed August 25, 2006)

CHAPTER 6 HOMEOWNERS ASSOCIATIONS AND CONDO OWNERS ASSOCIATIONS

Goldstein, Doris S. "Beyond the Homeowners Association: Blending Community Development Districts, Tax Exempt Organizations and Commercial POA's for Larger Planned Communities" Prepared for the Twenty-Eight Institute on Condominium and Cluster Developments October 30, 2003
Van Sickler, Michael and Zink, Janet. "CDD: The Hidden Cost of Living" St. Petersburg Times December 12, 2003

CHAPTER 7 PROPERTY TAXES

Orange County Government, Florida "Property Tax Questions and Answers" <http://orangecountyfl.net/cms/govern/pptytaxes/default.htm> (accessed September 9, 2006)

Bibliography

Volusia County Property Appraiser's Office "Our Appraisals and the Appeal Process" <http://webserver.vcgov.org/appeal.htm> (accessed March 13, 2006)
Duval County Property Appraiser's Office "Truth in Millage (TRIM) – Notice of Proposed Assessment" <http://www.coj.net/Departments/Property+Appraiser/Truth+in+Millage.htm> (accessed September 9, 2006)
Volusia County Property Appraiser's Office "State of Florida Homestead Exemptions" <http://webserver.vcgov.org/exmpt.htm> (accessed February 26, 2006)
Florida Governor Charlie Christ "Governor Christ Proposes Tax Cut Plan to Keep Florida's Economy Vibrant" <http://www.flgov.com/release/8567> (accessed February 4, 2007)

CHAPTER 8 PROPERTY INSURANCE

Florida Association of REALTORS "Special Edition: Legislators roll back Citizens premiums, insure vacation properties" *Florida Real Estate Headlines* Volume 12, Issue 3, January 23,2007
Freer, Jim "Will Citizens balloon into largest insurer in Florida?" *Orlando Business Journal* June 5, 2006 <http://Orlando.bizjournals.com/Orlando/stories/2006/06/05/story3.html> (accessed August 30, 2006)
Citizens Property Insurance Corporation "Who We Are" <http://www.citizensfla.com/help/WhoWeAre.htm> (accessed July 31, 2006)
Florida Governor Charlie Christ "Governor Christ Signs Bill Reducing Florida's Insurance Rates" <http://www.flgov.com/release/8551> (Accessed February 4, 2007)
Insure.com "Home Insurance Basics" <http://info.insure.com/home/basics.html> (accessed September 6, 2006)
The National Flood Insurance Program "What is Flood Insurance?" <http://www.floodsmart.gov/floodsmart/pages/whatfloodins.jsp> (accessed June 14, 2006)
The National Flood Insurance Program "Flood Zones Defined" <http://www.floodsmart.gov/floodsmart/pages/riskassesment/floodzonesdefined.jsp> (accessed September 16, 2006)

CHAPTER 9 CONTRACTS AND DISCLOSURES

Florida Association of REALTORS and The Florida Bar "Contract For Sale and Purchase" Rev. 7/04
Florida Association of REALTORS "Coastal Construction Control Line Disclosure" Rev. 7/06
Florida Department of Environmental Protection – Division of Beaches and Coastal Systems "The Homeowner's Guide to the Coastal Construction Control Line Program (Section 161.053, Florida Statutes)" <http://www.dep.state.fl.us/beaches/publications/pdf/propownr.pdf> (accessed August 30, 2006)

Florida Department of Environmental Protection – Division of Beaches and Coastal Systems "Frequently Asked Questions About the Coastal Construction Control Line" <http://www.dep.state.fl.us/beaches/data/pdf/cccl-faq.pdf> (accessed August 30, 2006)

The State of Florida Department of Community Affairs Codes and Standards Office "Florida Building Energy-Efficiency Rating System: New Residential Buildings"

Sheridan, Terry "Florida Construction Defects: Fewer claims, bigger checks" <http://www.oppenheimlaw.com/published_2004.html> (accessed August 28, 2006)

Florida Department of Business and Professional Regulation – Construction Industry Licensing Board "Frequently asked questions regarding the recovery fund for the Construction Industry Licensing Board" <http://www.myflorida.com/dbpr/pro/cilb/recov_faqs.pdf> (accessed September 19, 2006)

CHAPTER 12 REAL ESTATE FINANCE

National Reverse Mortgage Lender's Association "Reverse Mortgage Q & A" <http://www.reversemortgage.org/AboutReverseMortgages/ReverseMortgageQA/tabid/230/Default.aspx> (accessed February 11, 2006)

Cullen, Terri "Bridge Loans Offer Cash Infusion" RealEstateJournal.com: The Wall Street Journal Guide to Property <http://www.realestatejournal.com/buysell/mortgages/20030204-cullen.html> (accessed August 30, 2006)

Taylor, Don "Rate Locks Explained" Bankrate.com <http://www.bankrate.com/brm/news/DrDon/20020123a.asp> (accessed August 30, 2006)

Yahoo Finance "Good Faith Estimate: More Fees at Closing" <http://loan.yahoo.com/m/securing9.html> (accessed August 30, 2006)

Starker Services Inc. "Common Questions" <http://www.starker.com/faq.html> (accessed July 29, 2006)

"Check Your Score Before Getting a Loan" *Kiplinger's Retirement Report* June 2006: p. 5

CHAPTER 13 NEW CONSTRUCTION

Titan America "Frequently Asked Questions" <http://www.titanamerica.com/products/ready_mix/Virginia/faqs.html> (accessed September 6, 2006)

The Tile Roofing Institute "Why Tile?" <http://www.tileroofing.org/tileroofing/interior_ektid47.aspx> (accessed August 30, 2006)

Vandervort, Don "Metal Roofing: Benefits and Drawbacks" HomeTips.com <http://www.hometips.com/cs-protected/guides/metal_roofs/metalroof_benefits.html> (accessed August 30, 2006)

FLASH: Federal Alliance for Safe Homes "Garage Door: Securing" <http://www.flash.org/activity.cfm?currentPeril=1&activityID=112> (accessed September 2, 2006)

CHAPTER 16 REAL ESTATE CLOSINGS

American Land Title Association "Questions About Title Insurance" <http://www.alta.org/consumer/questions.cfm> (accessed October 17, 2006)

Equitable Title Agency, Inc. "Conventional Mortgage Rate Information" <http://www.equitabletitle.com/rates.html> (accessed October 17, 2006)

Florida Department of Revenue "Documentary Stamp Tax Collections" <http://www.myflorida.com/dor/taxes/doc_stamp_coll.html> (accessed September 9, 2006)

Kass, Benny L. "Housing Counsel: Protecting the Surviving Spouse" RealtyTimes.com June 20, 2005 <http://realtytimes.com/rtcpages/20050620_survivingspouse.htm> (accessed October 26, 2006)

Kass, Benny L. "Housing Counsel: Who is on Title to Inherited Property?" RealtyTimes.com July 24, 2006 <http://realtytimes.com/rtcpages/20060724_inheritedtitle.htm> (accessed October 26, 2006)

CHAPTER 18 FLORIDA RESOURCES

Wikipedia.org "Florida" <http://en.wikipedia.org/wiki/Florida> (accessed September 17, 2006)

Wikipedia.org "Tropical Cyclone" <http://en.wikipedia.org/wiki/Hurricane> (accessed September 4, 2006)

FLASH: Federal Alliance for Safe Homes "A Hurricane Overview" <http://www.flash.org/activity.cfm?currentPeril=1> (accessed September 2, 2006)

Enterprise Florida "Florida's Metropolitan Statistical Areas" June 2006 <http://www.eflorida.com/countyprofiles/pdfs/MSA_Indicators.pdf> (accessed September 14, 2006)

U.S. News and World Report "Best Hospitals 2006 – Florida" <http://www.usnews.com/servlets/HospSearch> (accessed August 16, 2006)

Airports Council International – North America "2005 North America Final Traffic Report: Total Passengers" <http://www.aci-na.org/asp/traffic.asp?art=125> (accessed September, 16, 2006)

VisitFlorida.com "Florida's Lighthouse Trail" <http://www.visitflorida.com/cms/d/floridas_lighthouse_trail.php> (accessed October 25, 2006)

the United States
00001B/79/A